LIBERAL RELIGION FOR EARTHLINGS

HOW TO BELONG TO A LIVING WORLD

J. CARL GREGG

CHAIR & TABLE PRESS

CONTENTS

Published by Chair & Table Press

Paperback ISBN: 979-8-9977256-0-0

ebook ISBN: 979-8-9977256-1-7

ALSO BY J. CARL GREGG

Forthcoming

Why Unitarian Universalism? A Living Tradition for a Changing World

For

Jim Edwards,
who showed me what liberal religion could be,

and

Jaylen,
who is growing up in a world that desperately needs
a liberal religion for Earthlings.

"We inter-are."[1]

—Thích Nhất Hạnh

INTRODUCTION

Do you really want to know?

I was not raised in a liberal religion for Earthlings. I grew up Southern Baptist in the Midlands of South Carolina, where I learned to think of this world as secondary, evolution as false, and heaven as home.

As a child, I would have been perfectly happy if every person on Earth were a Southern Baptist. My church taught me that the most important decision anyone could make was whether to accept Jesus as their personal Lord and Savior, and I believed it. Then I started meeting people who hadn't made that decision and who were unmistakably good people. Some of them were kinder, wiser, and more compassionate than many of the Christians I knew.

My best friend in middle school was an agnostic. In high school, I gravitated toward the drama club crowd, which meant that Brock, the only out gay kid in our school, became one of my friends. My church had taught me that there was something fundamentally wrong with him. The person I knew was funny, loyal, and braver than I was—not at all the person my church had taught me to fear. The more I learned about the world, and the more remarkable people I met beyond the boundaries of my childhood faith, the more my inherited assumptions began to unravel.

In college, as a religion and philosophy double major, I caught my first glimpses of unapologetically liberal religion, one that asked how to love this world and how to belong here. One of those moments remains especially vivid. After the end of my sophomore year, I was driving home when I stopped halfway at a bookstore. On a table of new books was John Shelby Spong's *Why Christianity Must Change or Die*. The title caught me. I picked up the book and started reading. It felt like an electric charge. New hardcovers were expensive, and I had very little spare money, but I knew I had to buy this one and would stay up that night until I finished it. I wouldn't necessarily recommend Spong's book today, but at that moment in my life, it helped fling open a door and gave me some of the courage I needed to walk through it.

Part of what struck me was the book's bold, direct declaration that the Christianity of my childhood needed to change—and that another kind of religion really was out there, one that could make room for the questions that had been opening inside me. Perhaps that is part of my hope for the title of this book as well: that someone might encounter the words *Liberal Religion for Earthlings* and wonder, Wait—is that a thing? You can be liberal and religious?

What kind of religion makes sense once we recognize that we are not at the center of the universe but on the third rock from the sun, orbiting an ordinary star near the edge of one spiral galaxy among hundreds of billions of others?

This book begins with the conviction that the scientific story of who we are changes how we understand ourselves and our place in the world. That conviction begins with a question Peter Mayer poses in the title of his song "Do You Really Want to Know?" In the lyrics, Galileo and Darwin each pause before revealing discoveries that would permanently unsettle humanity's understanding of itself. I discovered that I really did want to know. I still do.

There was no single moment when my old worldview collapsed. Rather, years of reading, study, relationships, and experience gradually taught me to let evidence lead and my beliefs follow, trusting that truth—even when it overturns my most cherished assumptions—is a gift to be welcomed rather than feared.

Yet knowledge alone does not transform us. Over the past five centuries, the Scientific Revolution has reshaped humanity's understanding of the universe and transformed civilization in countless ways, but it did not automatically make us wiser about how to live within the world it revealed.

Never before have we known so much about the universe and our place within it. Even so, we often continue to imagine ourselves as separate from the very world that sustains us. We celebrate individual achievement while forgetting our dependence on countless human and more-than-human relationships. We consume as though the Earth were merely a warehouse of resources rather than the larger web of life to which we belong.

Many religious and philosophical traditions have long taught that human life is bound up with the larger community of life. Modern science has shown just how literal those ties are. The question for a liberal religion for Earthlings is what that knowledge asks of us.

By *Earthling*, I mean more than someone who happens to live on Earth. I mean a creature formed by this planet's history, dependent on its living systems, and inseparable from the other lives it sustains. We are already Earthlings biologically. The task is to become Earthlings consciously and ethically: to recognize how deeply we belong and live from that knowledge.

Liberal religion names the approach through which I have learned to explore what that asks of us. For me, that path has led through progressive Christianity, Buddhism, and now Unitarian Universalism.[1] Here the word *liberal* is often misunderstood because it is heard primarily in partisan political terms. In this context, however, I am using the word in its older sense, from the Latin *liber*, meaning "free." Liberal religion (or free religion) is rooted in the conviction that truth need not fear honest inquiry.

Liberal forms of Judaism, Christianity, Islam, Hinduism, Buddhism, and many other religious traditions share this commitment. The liberal branches of different religions often have more in common with one another than they do with the more theologically conservative expressions of their own traditions. What unites them is not a shared creed or theology but a shared commitment to remaining open to wherever honest inquiry leads.

Rather than locating religious authority primarily in ancient texts, inherited dogmas, or institutional hierarchies, liberal religion believes that our understanding can—and should—evolve as we learn more about ourselves and the world. As Gary Dorrien writes in *The Making of American Liberal Theology*, liberal theology is "based on reason and experience, not external authorities."[2] In other words, don't believe something simply because someone tells you it is true. Believe it because it makes sense, because it is supported by the best available evidence, and because it is confirmed in your own direct experience. That commitment changes how we live.

Consider two questions that have divided religious communities in recent generations. Should women be ordained? Should same-sex couples be married? The answers depend on what one regards as the highest authority in matters of religion. Theologically conservative traditions have typically begun by asking what their ancient texts and inherited doctrines appear to teach. Liberal religion also values scripture and tradition, but believes they should be continually reinterpreted as human knowledge and experience grow. That willingness to learn from reality has led many liberal religious communities to ordain women and to marry same-sex couples.

Liberal religion has reshaped theology in many ways, from its understanding of scripture and religious authority to its commitments to democracy, human rights, gender equality, and LGBTQ+ inclusion. This book asks what happens when that same willingness to learn from reality is brought to bear on humanity's place within the living world.

Seen this way, the spiritual life is not about escaping the world but paying closer attention to it. Meaning, wonder, and transcendence are not waiting somewhere else. They are here: in this world, among the lives we share, and in the ordinary moments that make up our days.

The chapters that follow explore what that means in practice. They approach the larger inquiry from different angles.

Part I begins with ourselves. What changes when we see human beings as Earthlings: evolved animals, latecomers in a vast cosmic story, and participants rather than masters in the systems that sustain life?

Part II turns toward the more-than-human world. Trees, fungi, plants, and animals challenge many of the assumptions we have inherited about intelligence, perception, relationship, and our place among other forms of life.

Part III moves from understanding to practice. Attention, discernment, embodiment, and the rhythms of everyday life ask what it might mean to live more wisely within the world we are learning to see.

Part IV widens the horizon toward the future. If we belong to a living world, what does that belonging ask of us in the communities we build, the crises we face, and what we leave to generations we will never meet?

The movement is from seeing differently, to encountering differently, to living differently, to asking what our lives make possible for others.

How do we learn to belong to a living world?

PART I

BECOMING EARTHLINGS

Copernicus displaced Earth from the center of the cosmos. Darwin situated humans within the evolutionary tree of life. Ecology revealed the networks of dependence that sustain every living thing. Together, these discoveries challenged an inherited story: that human beings were specially created, set apart from the rest of nature, and given dominion over it.

As climate change and biodiversity loss accelerate, we are discovering how much our future depends on the health of the living world.

Part I begins close to home, with the practices through which a life takes shape, and then widens its view. It asks what changes when we see ourselves as evolved animals, relatives of other creatures, latecomers in a vast cosmic story, and participants in the systems that sustain life. Becoming Earthlings means learning to live from that knowledge.

1

TO DWELL POETICALLY

What does my life say I believe?

Growing up in South Carolina in the 1980s and 1990s, I rarely encountered outspoken atheists. Then, in college, I took a philosophy class from Jim Edwards, who had written this:

> I am a religious skeptic. I don't believe in Yahweh, Allah, Krishna, or Ogun; nor do I believe in the resurrection of bodies two days dead and buried, nor water changing into wine, nor men walking about unharmed in fiery furnaces.... About my skepticism, I feel neither pride nor shame; it's not an achievement, nor is it a failing. It's just a fact about me...a fact that doesn't have much to do with the way I actually live my life.[1]

I had never encountered atheism presented that way. Not as a rebellion or a wound, but as a considered conclusion someone could state plainly and without apology.

His comment eventually led me to a broader realization. Whatever we profess to believe, our lives are shaped less by our stated convictions than by our habits, our fears, and the social and economic conditions that form us.

Edwards helped me see how easy it is for thoughtful people to live good lives without religion. He also helped me see the reverse: that people could profess all the right beliefs and still fail to live well. I came to care less about what people said they believed than about what they did with their days.

If belief really drove behavior, this allegedly Christian country would have a smaller wealth gap and would distribute its resources according to need rather than perceived merit. The second chapter of Acts describes members of the early church selling their possessions and distributing the proceeds "to all, as any had need." I have attended a great many churches. I have never attended that one.

Edwards's suspicion also ran in both directions. He was skeptical not only of traditional religious belief but also of the secular temptation to exchange one absolute authority for another. He resisted making idols of "Humanity," "Science," or "Reason." Even our highest ideals can become dangerous when they are treated as unquestionable authorities rather than as tools for helping us live wisely.

What ultimately mattered to him were the ordinary practices of honesty, care, responsibility, and attention. In his words,

> If I can be said to 'believe in' anything at all, it's a series of fairly small things: I believe in keeping the Kirtland's Warblers alive, in keeping (most of) one's promises, in cultivating a decent regard for the truth, in offering kindnesses to strangers (including the strangers within ourselves), in making things better for the sick and the poor, in preserving a healthy vulgarity, and so forth.[2]

Passages like these give some sense of why Edwards was such an electrifying teacher. On a fairly religiously and politically conservative campus, students were drawn to his classes even when they expected to disagree with him. At a liberal arts college that prized teaching, he became the first professor to receive its Meritorious Teaching Award twice. What drew us, I think, was his intellectual honesty. He did not flatter us or soften what he thought to be true. He spoke with the candor of someone who had spent decades asking hard questions and was willing to tell us, as plainly as he could, where that search had led.

That willingness to follow difficult questions wherever they led made me especially curious when, during my sophomore year, Edwards, the campus atheist, offered a philosophy of religion course based on his newly published book, *The Plain Sense of Things: The Fate of Religion in an Age of Normal Nihilism*. I had expected him to have little interest in religion. Instead, one of the book's central questions is what religion could still mean in the wake of Copernicus, Darwin, Freud, Einstein, and other paradigm-shifting thinkers. His answer—"to dwell poetically on the earth as a mortal"—has influenced my own understanding of what religion can be.[3] I have been chewing on that phrase for two decades. It lies at the heart of the kind of spirituality this book seeks to explore: learning to belong to a living world.

When Edwards asks, "What should it mean for us to be religious?" and answers, "to dwell poetically on the earth as a mortal," he is using *poetically* in a broad and deep sense, rooted in the ancient Greek *poiesis*: "to make." In this older sense, poetry is more than writing verses or expressing emotions. It is about creative activity: bringing something into being. At its heart, it is about shaping how we live in the world. Edwards understood religion in those terms—not primarily as a system of beliefs but as a way of making a life.[4]

If religion is about shaping human lives, then its central task is not merely to teach doctrines but to develop practices that form our character. Even when the religious stories that once sustained those practices are no longer credible when taken literally, Edwards believed the practices themselves could still nurture honesty, awareness, and ethical responsibility:

> We need and luckily still have available to us practices that can contain, concentrate, and transmit the sacramental energies—energies for limitation in the face of hubris and for transformation in the face of complacency—that used to be bound up in the stories of the gods.[5]

In plainer terms, we need practices that check us when we grow arrogant and unsettle us when we grow complacent. The forms those practices take have changed over time. The need for them has not. For Edwards, the enduring value of religion lay less in supernatural beliefs than in the ways of living that form the kind of people we become.

Practices matter because what we do is almost always more important than what we say we believe. Social psychologists have a phrase for this: behavior is believable. So the real question is which of those practices can still shape a life.

What forms of religious life remain once we refuse to choose between intellectual honesty and spiritual depth? Some of them look like poetry and art. But it would be exhausting to try to be spectacularly creative all the time. We also need ordinary habits that quietly form us.

One story illustrates this approach. Albert Michelson, the first American to receive a Nobel Prize in the sciences, was a talented billiards player. When someone commented on his skill, he once replied that billiards was a fine game, but that chess was better. He added that painting was better still—and that physics surpassed them all. He then put away his cue, walked home, and spent the afternoon painting beneath a large tree in his front yard.[6]

My own version would run: Ultimate Frisbee is a good game. But Frisbee is not as good a game as watching and arguing about films. And film criticism is not as good a game as preaching, teaching, and writing books. Then I might go home and watch a film.

The ranking is not the point. What stays with me is the spaciousness of a life in which even one's highest calling does not crowd out every other delight.

So I will ask you what Edwards's question asks me: What are your practices of authenticity? What makes you feel grateful, more alive, more yourself, more connected to other people and to the world? And what would you need less of to make room for it?

To dwell poetically on the earth does not require a dramatic or spectacular life. More often, it means learning to inhabit ordinary life with greater awareness and gratitude: to discover beauty, fragility, and interdependence in experiences we might otherwise rush past. How we attend to the ordinary moments of our lives matters because, for the most part, those ordinary moments become our lives. As Annie Dillard writes, "How we spend our days is, of course, how we spend our lives."[7]

Henry David Thoreau offers one of the clearest examples of this way of living. In *Walden*, even ordinary life becomes a form of spiritual practice: "I got up early and bathed in the pond; that was a religious exercise."[8] For those of us who can turn on a cascade of hot water whenever we want, even the morning shower can become an occasion for gratitude, savoring, and refreshment rather than merely one more thing to finish before we get out the door. The same is true of walking the dog, cooking dinner, and caring for a child. To dwell poetically is to bring that spirit of gratitude and mindfulness into ordinary life—not by escaping the everyday, but by inhabiting it more fully.

Dwelling poetically also requires receptivity. We can practice attention and gratitude, but we cannot manufacture everything that gives life meaning. Edwards eventually reaches for the language of grace:

> I cannot survey the contents of my imagination as I can the contents of my sock-drawer. What appears in my imagination appears spontaneously, as unpredictably as a river

> floodtide in spring. I can ask for water, but I cannot make it rain. I can work and work on my lines, but I cannot add a cubit to the stature of even a single sentence. That, if it comes, is grace.[9]

Grace, in this sense, is not supernatural intervention. It is what we call those gifts (a flash of inspiration, the insight you didn't work for) that we can prepare ourselves to receive but cannot command.

The final word in Edwards's phrase is *mortal.* To dwell poetically on the earth is to do so without denying that our time here is limited. In a lecture titled "What Really Matters?" he concluded with words from *Walden*: "Be it life or death, we crave only reality. If we are really dying, let us hear the rattle in our throats and feel cold in the extremities; if we are alive, let us go about our business."[10]

He ended his final book, *The Plain Sense of Things: The Fate of Religion in an Age of Normal Nihilism*, by returning to the question that had shaped so much of his thinking:

> What should it mean for us to be religious? I have answered it thus: to dwell poetically on the earth as a mortal. That the question and its proper answer might at some time be unintelligible or trivial is not the point, so far as I can see. This is our life: Should we not live it as simply and sincerely and joyfully as we can?[11]

2

INSIGNIFICANT ANIMALS, DISSATISFIED GODS

What kind of animal are we?

Yuval Noah Harari opens *Sapiens: A Brief History of Humankind* by asking us to take a walk:

On a hike in East Africa 2 million years ago, you might well have encountered anxious mothers cuddling their babies and clutches of carefree children playing in the mud.... There was nothing special about humans. Nobody, least of all humans themselves, had any inkling that their descendants would one day walk on the moon, split the atom, fathom the genetic code and write history books. The most important thing to know about prehistoric humans is that they were insignificant animals with no more impact on their environment than gorillas, fireflies, or jellyfish.[1]

Insignificant animals. That is where we start.

But over immense stretches of time, one small branch of the evolutionary tree acquired the power to alter the fate of an entire planet. Around 150,000 years ago, the total human population may have numbered only about one million. Today, there are more than eight billion of us, and our species has become capable of reshaping the Earth itself—for good and for ill. The species that learned to read its own genome is now unsettling the climate that produced it.

How did one ordinary primate come to acquire such extraordinary power? You've seen the T-shirt: an ape on the left, then an ape standing straighter, then a caveman, then a caveman with better posture, and finally a modern human striding off the right edge. The picture is wrong. Evolution is not a ladder; it is a tangled, branching tree.

Human beings did not emerge alone or all at once. Members of the genus *Homo* evolved in East Africa roughly 2.5 million years ago, and for most of our evolutionary history multiple human species coexisted. As recently as about 50,000 years ago, *Homo sapiens* shared the planet in different regions with Neanderthals, Denisovans, and *Homo floresiensis*, along with other archaic human lineages that remain imperfectly understood. Our own lineage is just one branch of that evolutionary tree, shaped by contingency, adaptation, extinction, and unexpected turns. Today, *Homo sapiens sapiens* is the dominant species on Earth, but this is a relatively recent development within the planet's 4.5-billion-year history.[2]

Remembering this evolutionary story can loosen the illusion that human beings somehow stand outside or above the rest of the living world. That shift in perspective is one of the gifts liberal religion can offer: allowing our spiritual imagination to be reshaped by what we have learned through science and experience. We are not the predetermined culmination of creation, but one fragile branch within a much older and vastly more complex web of life.

It also helps explain how we came to occupy such an unusual place in the history of life. For most of evolutionary history, human beings occupied the middle of the food chain. We hunted smaller animals, while larger predators sometimes hunted us. But through innovations such as the controlled use of fire, increasingly sophisticated tools, cooperation, and language, our species rose to ecological dominance with astonishing speed.

Unlike lions or sharks, which evolved gradually into the role of apex predators over millions of years, human beings acquired extraordinary power in only a geological instant. Ecosystems had little time to adapt to our emergence—and, in many ways, neither did we. We possess the technological power to reshape the planet, but we still carry the emotional instincts of a species that evolved under conditions of scarcity, vulnerability, and immediate danger. The result is extraordinary creativity and cooperation—but also destruction on an unprecedented scale.[3]

Recognizing this tension may be one of the first steps toward inhabiting the Earth more wisely and humbly. Our extraordinary power comes not from biological evolution alone, but from our ability to transmit stories, habits, knowledge, and institutions across generations. Alongside biological evolution, there is cultural evolution.

Whereas genes are inherited biologically, cultural

patterns are learned and transmitted across generations. Harari calls these learned patterns "artificial instincts."[4]

Richard Dawkins offers a related way of understanding this process. He coined the term *meme* from the Greek word *mimēma*, meaning "that which is imitated," deliberately shortening it to echo the word *gene*.[5] Memes spread through teaching, storytelling, and other forms of cultural transmission. Religions, national identities, prejudices: none of these are in our DNA. They are taught.

Consider a beehive. Within it we find intricate patterns of cooperation, labor, communication, and social organization. But much of this behavior is instinctive and genetically programmed. As Harari memorably observes, "Bees don't need lawyers, because there is no danger that they might forget or violate the hive constitution."[6]

Human societies work differently. Laws, customs, and manners have to be taught to every new generation, or they are gone in one. Unlike bees, we can reject what we inherit culturally, reinterpret it, or throw it out.

That cultural flexibility is both one of humanity's greatest strengths and one of its greatest dangers. It allows us to revise inherited norms and build cultures that nurture compassion, cooperation, and justice. It also allows systems of domination, violence, greed, tribalism, and ecological destruction to be taught and sustained across generations.

In the Declaration of Independence, Thomas Jefferson wrote: "We hold these truths to be self-evident, that all men are created equal, that they are endowed by their Creator with certain unalienable Rights, that among these are Life, Liberty and the pursuit of Happiness." Jefferson writes as though these truths were built into reality itself. Harari suggests that an honest biologist would have to rewrite the sentence: "We hold these truths to be self-evident, that all humans evolved differently, that they are born with certain mutable characteristics, and that among these are life and the pursuit of pleasure."[7] Nobody would fight a revolution over that sentence.

Jefferson was not merely choosing elevated rhetoric: like many Enlightenment thinkers, he believed that human equality and rights were grounded in natural law. Harari's point is that biology alone cannot establish those moral claims.

The same distinction applies to the 1948 Universal Declaration of Human Rights. Rather than describing what evolution inevitably produces, it articulates a moral commitment, a vision of the kind of world human beings can choose to create and sustain together. That does not make human rights less meaningful. It makes them more precious. Ideas such as equality, dignity, democracy, pluralism, and universal compassion are not inevitable features of the universe. They are fragile cultural achievements that human beings have struggled to imagine, realize, preserve, and expand across generations—and that each generation must choose to sustain.[8]

Just as biological evolution does not guarantee democracy or human rights, cultural evolution does not guarantee that we will choose wisely enough to preserve the conditions under which our own species will continue to flourish. The ideas we pass from one generation to the next may now matter as much as the genes we pass on. Those cultural choices now carry planetary consequences.

Earth's history does not repeat itself exactly, but certain

patterns do recur. Since Earth has already undergone five previous mass extinctions over the past half-billion years, we should take seriously the scientific warning that human activity may now be driving a sixth.[9]

Harari imagines one possible future if we continue down our current path:

> Rats and cockroaches, for example, are in their heyday. These tenacious creatures would probably creep out from beneath the smoking rubble of a nuclear Armageddon, ready and able to spread their DNA. Perhaps 65 million years from now, intelligent rats will look back gratefully on the decimation wrought by humankind, just as we today can thank that dinosaur-busting asteroid.[10]

That is Harari's consolation, and it is a cold one. Life goes on; it simply may not go on with us, or with much that we love. But whether it does may be up to us.

Harari ends *Sapiens* by asking what our species will do with the extraordinary powers it has acquired: "Is there anything more dangerous than dissatisfied and irresponsible gods who don't know what they want?"[11] Some of us know what we want. That is what the rest of this book is about.

If evolution explains how we became the kind of creatures we are, culture helps determine the kinds of creatures we can become. Biology has brought us this far. The future now depends, in no small measure, on the stories we tell, the communities we build, the virtues we learn, and the responsibilities we choose to embrace. What is at stake is not only who each of us becomes, but what we become together.

Meeting that challenge requires more than scientific

understanding. It also calls for moral and spiritual formation—the kind liberal religion has long practiced alongside many other religious, philosophical, and ethical traditions. Across history, these traditions have sought, in very different ways and with varying degrees of success, to teach compassion over cruelty, truth over propaganda, and cooperation over tribalism. At their best, they challenge the old illusion that we stand apart from the larger community of life.

3

YOUR DOG REALLY DOES LOVE YOU

Where does my kinship stop?

Twenty-five hundred years ago, the Greek philosopher Protagoras declared that "man is the measure of all things."[1] It is understandable that we would say so—surprise!—about ourselves. The claim became harder to defend after Darwin moved us from a little lower than the angels to a little higher than the apes.[2] The belief survives anyway: that human beings are fundamentally unlike every other creature on Earth. Philosophers sometimes call this human exceptionalism.

But the idea is not merely philosophical. Human influence on the planet has grown so large that many scientists now propose the term Anthropocene (the age of humans) for a new geological epoch, one in which human activity has become one of the dominant forces shaping the Earth.

As Elizabeth Kolbert argues in *The Sixth Extinction: An Unnatural History*, human exceptionalism has become one of history's great ironies: "In pushing other species to extinction, humanity is busy sawing off the limb on which it perches."[3] The more power we acquire over the natural world, the more clearly we reveal that we have never stood apart from it.

Moving beyond human exceptionalism requires holding two truths together. First, human beings are not as exceptional as earlier generations imagined. Second, the other beings with whom we share this planet matter far more than previous generations often recognized. They matter in their own right, and we depend on them, directly and through the ecosystems they sustain, for the conditions that make human life possible.

For much of Western history, influential thinkers assumed that human beings alone possessed the capacities that mattered most, shaping how generations understood moral worth and humanity's place in the natural world. The seventeenth-century French philosopher René Descartes made the claim especially starkly, arguing that animals lacked reason and genuine thought and that their behavior could be explained mechanically.

More than a century later, Voltaire was having none of it. He wrote about Descartes's position with the same contempt Descartes had shown animals:

> What a pitiful, what a sorry thing to have said that animals are machines bereft of understanding and feelings.... Bring the same judgment to bear on this dog which has lost its

> master, which has sought him on every road with sorrowful cries, which enters the house agitated, uneasy, which goes up the stairs, from room to room, which at last finds in his study the master it loves, and which shows him its joy by its cries of delight, by its leaps, by its caresses.[4]

If you have lived with animals, you have your own version of this. My dogs each had one bark for wanting to be fed, another for wanting out, and a third for the fearsome postal worker. My cat, when I am late with dinner, looks me dead in the eyes and knocks over her bowl. She is not confused about what she is doing.

Darwin gave scientific expression to that intuition. In *The Descent of Man*, he argued that the differences between humans and other animals are often differences of degree rather than kind. Such a conclusion challenged one of humanity's oldest assumptions: that we occupy an entirely separate category of being. As Darwin observed,

> Everyone has heard of the dog suffering under vivisection, who licked the hand of the operator. This man, unless the operation was fully justified by an increase of our knowledge, or unless he had a heart of stone, must have felt remorse to the last hour of his life.... Animals, whom we have made our slaves, we do not like to consider our equal.[5]

Darwin's insight becomes even more striking in light of evolution itself. All life on Earth is related. Humans and chimpanzees diverged from a common ancestor roughly five to six million years ago, and our two genomes differ by only about 1.2%.[6]

Even species separated from us by hundreds of millions of years remain connected through the branching evolutionary tree of life. As the philosopher of science Peter Godfrey-Smith explains in *Other Minds: The Octopus, the Sea, and the Deep Origins of Consciousness*, for any two living animals, "we can trace two lines of descent down the tree until they meet in a common ancestor."[7]

Octopuses, squid, and cuttlefish are astonishingly intelligent. Humans and octopuses last shared a common ancestor roughly 600 million years ago, and evolution independently gave rise to complex nervous systems and sophisticated behavior in both lineages.[8] Cephalopods represent "an independent experiment in the evolution of large brains and complex behavior."[9]

The independent evolution of complex intelligence raises difficult questions about consciousness. The word itself can mean several things. Sometimes we use it to mean reflective self-awareness: the ability to think about our own thoughts and actions. At other times, we mean a heightened state of awareness, as when danger sharpens every sensation. But in its most basic sense, consciousness means that experience is happening from the inside: that there is something it feels like to be a particular creature.

Start with something that shows no sign of consciousness. The pulpit I preach behind on Sunday mornings gives us no reason to think it has an inner life. I could chop it up and burn it, and it would make no difference to the pulpit. A pulpit can be damaged, but nothing feels that damage from the pulpit's point of view.[10] A conscious being, by contrast, has an inner life.

Once we ask which creatures have an inner life, the evidence suggests that consciousness and complex cognition are far more widespread than we once imagined:

- Chickens live in stable social groups, recognize each other by their faces, and have twenty-four distinct cries, including separate alarm calls for different threats.[11]
- Sheep can recognize and remember the faces of at least fifty other sheep for more than two years, and they can also distinguish familiar human faces.[12]
- Jays appear capable of planning ahead for future needs.[13]
- Crows and parrots demonstrate remarkable problem-solving abilities.[14]
- Gorillas and capuchin monkeys fashion and use tools.[15]
- Octopuses navigate mazes, unscrew jars, and distinguish between individual human caretakers.[16]
- Elephants grieve. They return to the bones of dead family members, gently explore them with their trunks, and sometimes stand vigil beside them for hours.[17]

Taken together, these examples point to emotional and cognitive systems that are evolutionarily ancient and widely shared. As the marine ecologist Carl Safina writes in *Beyond Words: What Animals Think and Feel*, hormones connected to bonding and attachment, including oxytocin and vasopressin, predate the divergence of many major evolutionary branches by hundreds of millions of years. Safina summarizes the implications simply: "Your dog really does love you."[18]

The science leads directly to an ethical question. As the philosopher Peter Singer has argued, if other beings can suffer, on what basis should moral concern stop at our own species?[19]

Part of the answer lies in our shared evolutionary history. The capacities to suffer, form bonds, care for offspring, seek safety, and experience fear did not begin with human beings. We are animals ourselves, and these capacities arise from ancient biological systems shared across many lineages. Evolutionary distance does not erase those continuities.

A human being is not an octopus, a chicken, a chimpanzee, or a dog. The differences are real, but they do not erase our kinship. Where exactly moral concern should stop is genuinely hard to say. Some creatures give us much clearer evidence of suffering than others, and none of us can live without affecting other life. Kinship asks us to pay attention to what other creatures may feel and to reduce needless suffering where we can.

So may we each do what we can, within our spheres of influence, to decrease suffering and cruelty and to increase freedom and compassion for all sentient beings.

The Buddhist practice of *metta* ("loving-kindness") gives these aspirations a form. Before you turn the page, try reading these intentions aloud, slowly and with an open heart—and notice what you feel:

May all sentient beings be filled with loving-kindness.

May all sentient beings be well.

May all sentient beings be peaceful and at ease.

May all sentient beings be free.

4

LOOK AND SEE

How do I know?

As a child, I remember realizing that there was no obvious stopping place for the question "Why?" It felt as though the floor had dropped away beneath me. It was both unsettling and exhilarating to discover that every answer opened onto another question.

Now, years later, raising my own young son, I am experiencing that same endless chain of questions from the other side.

"It's time for bed."

"Why?"

"Because if you don't go to bed, you'll be tired in the morning."

"Why?"

"Because your body needs sleep."

"Why?"

"Because that's how our bodies work."

"Why?"

"Because that's how humans evolved."

"Why?"

Parents develop various strategies for responding to this relentless interrogation. The classic answer is, "Because I said so." A more philosophically inclined parent might simply shrug and reply, "Why not?" One of my favorites is to turn the question back on my son: "Why do *you* think?" There's also the industrious approach: "Let's go to the library tomorrow and find out together." Or perhaps the quintessential twenty-first-century response: "I don't know. Here's my phone. Ask AI."

That childhood discovery of the power of "why" has stayed with me. I remain suspicious of answers that bring questioning to a premature end. But I also came to see that questions are not neutral. They carry assumptions about what kind of answer the world can give. The task is not merely to keep asking why, but to notice what the question already assumes—and test that assumption against reality.

Asking better questions can reveal that some of the ways we organize our lives, our institutions, and our societies persist not because they are well founded, but because they have gone unquestioned. Far from being an exercise in endless regression, persistent questioning has often opened new ways of seeing the world. Again and again, some of humanity's greatest intellectual and scientific breakthroughs have begun when someone refused to stop asking why.

If you keep following those questions long enough, you eventually arrive at some of humanity's biggest ones. The

John Templeton Foundation once invited twelve prominent scientists, philosophers, theologians, and public intellectuals to answer, in five hundred words or fewer, a deceptively simple question: "Does the universe have a purpose?" Their answers covered nearly every possibility: "Unlikely. Very likely. Yes. Yes. Yes. Certainly. No. No. Not sure. Perhaps. Indeed. I hope so."[1]

Of all the responses, the one I find most intriguing comes from the astrophysicist Neil deGrasse Tyson. His answer begins with three refreshingly honest words: "I'm not sure." He argues that anyone claiming certainty about the universe's ultimate purpose is asserting knowledge that goes beyond what the available evidence can establish. Rather than beginning with what we hope is true, Tyson asks us to begin with the universe as we actually find it.

From that perspective, he observes that "humans were not around to ask these questions for 99.9999% of cosmic history. So if the purpose of the universe was to create humans then the cosmos was embarrassingly inefficient about it." He likewise notes that a universe supposedly designed as "a fertile cradle for life" is also one in which volcanoes, earthquakes, storms, pestilence, and "especially killer asteroids have left extinct 99.9% of all species that have ever lived here."

Then he turns the question around. If you are religious, you might say the purpose of life is to serve God. But if you were one of the hundred billion bacteria in a single cubic centimeter of your lower intestine, you would say the purpose of human life is to provide you with a dark but idyllic anaerobic habitat of fecal matter. Tyson's point is not that the universe is meaningless. It is that we should be cautious about claiming to know whether it was created for a particular purpose.

That same willingness to reckon honestly with what we find animates journalist Jim Holt's exploration of a related question in his book *Why Does the World Exist?: An Existential Detective Story*. He jokes that the universe was created by "a being that is 100% malevolent, but only 80% effective."[2] My worldview is not that pessimistic. But the line stayed with me because any honest account of existence must reckon with both the beauty of the world and the suffering within it.

That kind of questioning requires a willingness to examine our assumptions and allow reality to challenge the answers we have inherited. As we grow older, however, we sometimes lose that spirit of inquiry, becoming more interested in defending familiar answers than in following the evidence wherever it leads. That contrast reminds me of Christopher Durang's satirical play *Sister Mary Ignatius Explains It All for You*, in which a nun quizzes a seven-year-old boy on the memorized answers of Catholic catechism:

> Q. Who made the world?
> A. God made the world.
> Q. Who is God?
> A. God is the Creator of heaven and earth, and of all things.
> Q. What is man?
> A. Man is a creature composed of body and soul, and made to the image and likeness of God.
> Q. Why did God make you?
> A. God made me to know Him, to love Him, and to

> serve Him in this world, and to be happy with Him for ever in heaven.

Unlike the child's persistent "Why?" these questions are not meant to explore reality but to reinforce certainty.

Behind that catechism stands one of the foundational stories of Western civilization. The opening verses of Genesis read:

> When God began to create the heavens and the earth, the earth was complete chaos, and darkness covered the face of the deep, while a wind from God swept over the face of the waters. Then God said, "Let there be light," and there was light.[3]

This biblical creation account is commonly understood as depicting God creating the universe out of nothing. But a closer reading suggests a more nuanced picture. Before God speaks, there are already darkness, "the deep," and the waters over which God's Spirit hovers.

My point here is not to settle how Genesis should be interpreted. It is to notice something more basic: the stories we inherit help determine which questions occur to us in the first place. If we assume the universe was intentionally created, it is natural to ask what purpose its Creator had in mind. If we assume it developed through natural processes instead, the question shifts: we are more likely to ask how it came to exist than why.

The influence of those starting assumptions can be seen throughout the history of Christian thought. For centuries, many Christians understood the opening chapters of Genesis to provide a literal chronology of the universe. One of the best-known examples is Anglican Archbishop James Ussher, who in 1650 calculated the date of creation as October 23, 4004 B.C.E., making the universe only about six thousand years old. That once-common assumption has become increasingly untenable in light of modern cosmology.

Today, many people begin not with ancient scripture but with the story modern cosmology tells about the universe. That story, too, reshapes which questions seem natural, and which questions may no longer make sense in quite the same way.

Holt explores one intriguing response to the question of what came before the Big Bang. If time itself begins with the universe, there would be no moment "before" $t = 0$ because time itself would be part of what comes into existence. In that sense, although the universe is finite in age, it has "always" existed, if by "always" we mean every moment of time.[4]

But that is not the only possibility. If we trace the universe backward using our current theories, we eventually reach conditions where those theories can no longer tell us confidently what happened. There may have been no "before" in any meaningful sense, or the answer may require a deeper account of reality than we yet possess. In either case, the question itself reminds us to examine the assumptions built into even our most natural ways of asking.

The more closely I look at what modern science reveals, the more mysterious reality becomes. That sense of mystery has grown as our picture of the cosmos has expanded. As recently as the early twentieth century, many astronomers still believed that the Milky Way was the universe. Today, astronomers estimate that the Milky Way is one galaxy among hundreds of billions, perhaps as many as two trillion, in the observable universe.[5] Some physicists even speculate that our universe is one of many—a multiverse.[6] Each new discovery not only answers old questions but also changes the questions we think to ask.

The question is not only what we discover about the universe, but how those discoveries shape the way we see ourselves. The vastness of space and time can inspire despair at our seeming insignificance. But it can also inspire wonder. There *is* something rather than nothing. We are here, awake to this astonishing universe, sharing this brief moment with one another. That is no small thing.

To live wisely within such a vast universe, we need the humility to let reality surprise us. That openness to discovery is, for me, one of liberal religion's greatest gifts: the willingness to examine the stories and assumptions we inherit rather than treat them as beyond question.

The philosopher of science Thomas Kuhn used the word *paradigm* for the shared assumptions, models, and practices that guide scientific work. We usually see the world through a paradigm without noticing the paradigm itself. The temptation is to mistake it for reality. Observations that resist the prevailing framework can expose its limits and point toward a better explanation.[7]

As the philosopher Ludwig Wittgenstein put it, "Don't say, 'There must be...' Look and see."[8] Again and again, looking has overturned what once seemed obvious. Good thinking begins with careful attention to reality.

Left to ordinary intuition, it can seem obvious that our planet is the center of the universe. Galileo looked through a telescope and saw a sky that did not revolve around us.

Left to ordinary intuition, it can seem obvious that our species is the pinnacle of creation. Darwin didn't just think. He spent five years on a ship and years more with barnacles, and saw humanity as one twig on the tree of life.

The more I learn about the universe, the less convinced I am that reality fits neatly within any single explanation. Rather than assuming, like Sister Mary Ignatius, that we can explain everything, I find myself increasingly grateful both for what human inquiry has revealed and for the mysteries that remain.

The history of science reminds us that reality has often turned out to be stranger and larger than previous generations imagined. If we learn to live sustainably on this planet, future generations will almost certainly discover aspects of reality we cannot yet conceive of. What they take for granted may seem as extraordinary to us as galaxies, black holes, and quantum fields once seemed to our ancestors.

Why does the world exist? Why is there something instead of nothing? Why? Those questions may never leave us. But honest inquiry requires more than repeating them. We must notice what they assume, test those assumptions against the world, and allow reality to revise not only our answers but our questions. Don't just think. Look and see.

5

WHAT DOESN'T GO AWAY

What am I willing to let reality correct?

As a teenager, I sometimes attended events with my theologically conservative Christian youth group where T-shirts were sold bearing slogans such as "I believe in the Big Bang. God spoke and—*bang*—it was." Beneath the humor was a deeper posture toward truth: the conviction that the most important truths had already been revealed long ago, and that religion's task was to defend those inherited certainties against new understandings of the world.

For much of Christian history, scripture and tradition served as primary authorities for claims about the natural world. The old slogan captured that approach: "God said it. I believe it. That settles it." The Creation Museum in Kentucky, however, presents young-Earth creationism as scientific history, complete with Adam and Eve among the dinosaurs. Science has changed our culture so completely that even many of its critics now speak its language.

Science gained that authority because it worked. Ideas increasingly came to be judged by how well they matched reality rather than by who had first proposed them. Science predicted eclipses, cured infections, and put people on the moon. It also gave us the machine gun and the atomic bomb.

This new way of understanding did not emerge fully formed. Its rise represented one of the great intellectual transformations in human history. In *To Explain the World: The Discovery of Modern Science*, physicist Steven Weinberg traces how humanity gradually developed the methods and habits of mind that became modern science.

The emergence of modern science required more than curiosity or intelligence. It required a willingness to let reality correct our assumptions. Many early philosophers asked profound questions about reality, but their answers often depended more on logical reasoning and speculation than on systematic observation and experiment.[1]

The Pre-Socratic philosophers went looking for the one substance underlying everything. Thales said water. Anaximander proposed an indefinite, boundless source from which everything arose. Anaximenes said air. Heraclitus said fire, and Empedocles proposed earth, air, fire, and water in combination. Every one of these is interesting. But they are speculative starting points, not experimentally grounded theories. You cannot start from earth, air, fire, and water and eventually arrive at an iPhone.[2]

A similar pattern appeared in theology. In earlier eras, religious arguments were often shaped less by evidence than by symbolic correspondence. The second-century theologian Irenaeus explained why there are four Gospels and not more: "The Gospels could not possibly be either more or less in number than they are. Since there are four zones of the world in which we live, and four principal winds...the pillar and foundation of the Church is the gospel...it fittingly has four pillars."[3] The same style of reasoning could just as easily justify three Gospels or five. That is what makes it sophistry: clever reasoning that collapses the moment you step back from it.

Today, many readers will find such arguments strange. They remind us that many premodern thinkers approached truth differently. The question was often not, "What does the evidence show?" but, "What pattern best reveals God's design?" Such reasoning rarely generated the kind of testable predictions that made knowledge cumulative.[4]

One of the earliest bridges to modern scientific thinking emerged through something as seemingly intangible as music. The Pythagoreans discovered that harmonious musical intervals corresponded to simple mathematical ratios:

Octave—2:1
Fifth—3:2
Fourth—4:3

This meant that something as beautiful and subjective as music was governed by objective, discoverable patterns. This remarkable connection between mathematics and music suggested that hidden numerical patterns could help explain the world. More significant still, it hinted that patterns in nature could be observed, measured, tested, and used to make reliable predictions.[5]

At the same time, mathematics alone is not enough. It is not itself a natural science but a language for describing and analyzing the physical world. Mathematical theorems are established through logical proof; scientific theories stand or fall by how well they accord with observation and experiment.[6]

Modern science emerged from the partnership between mathematical reasoning and empirical evidence. Elegant ideas were tested against reality through observation, measurement, and experiment. Over time, shared methods allowed scientific knowledge to accumulate as claims were tested, corrected, refined, and extended.[7]

Careful study, of course, was nothing new. Human beings had long observed the world around them. More than five thousand years ago, ancient Egyptians tracked the stars closely enough to predict the seasonal flooding of the Nile River.[8]

Likewise, it is a myth that everyone thought the world was flat before Columbus. Greek philosophers had argued for a spherical Earth 2,500 years ago, from the curve of Earth's shadow on the Moon during a lunar eclipse and the way the stars shift as you travel north or south.[9]

The argument in Columbus's day was not about the shape of the Earth but its size. He thought it small enough to sail from Spain to Asia without running out of food and water. He was wrong. He reached land before his provisions ran out only because the Americas lay between Europe and Asia.[10]

These examples show that scientific thinking did not

emerge all at once. It developed slowly as humanity became increasingly willing to let careful observation of the natural world challenge appearances, assumptions, and inherited beliefs.

The decisive transformation, however, began in the sixteenth century with the Scientific Revolution. At its symbolic center stood Nicolaus Copernicus. His work challenged one of humanity's deepest assumptions: that the Earth occupies the center of the cosmos.[11]

His *On the Revolutions of the Heavenly Spheres* was published in 1543, just twenty-six years after Martin Luther posted his "Ninety-Five Theses." The Protestant Reformation and the Scientific Revolution thus emerged at nearly the same time. Ironically, even as many Protestant reformers challenged the authority of the Catholic Church, most remained resistant to the emerging scientific worldview.[12] Luther himself dismissed Copernicus sharply, writing: "This is what that fool does who wishes to turn the whole of astronomy upside down.... I believe in the Holy Scriptures, for Joshua commanded the Sun to stand still and not the Earth."[13]

Here the growing tension between scriptural authority and empirical observation came clearly into view. Traditional interpretations of sacred texts often appeared to conflict directly with conclusions drawn from observation, mathematics, and the emerging scientific method. When traditional religious beliefs and empirical evidence point in different directions, which should carry the greatest authority?

Defenders of the new scientific worldview could be combative as well. Long before the rise of the "New Atheist" movement, astronomers and natural philosophers sometimes responded to religious resistance with open ridicule. Johannes Kepler included in one of his books a section titled "Advice for idiots." His advice: anyone too stupid to understand astronomy, or too weak to believe Copernicus without it affecting his faith, should give up philosophy altogether and "betake himself home to scratch in his own dirt patch."[14]

Ridicule is rarely the most effective way to change minds, but the broader historical pattern remains important. Human beings often resist discoveries that threaten their worldview, social standing, or economic interests.

Resistance to heliocentrism shares important similarities with more recent forms of scientific denialism surrounding evolution and climate change. The same sophistry reappears whenever identity or ideology takes precedence over evidence. In 1616 the Church's *Index of Forbidden Books* suspended *On the Revolutions* until it could be corrected to present heliocentrism as a hypothesis rather than fact. The reasoning that put it there has not gone extinct.

The same principle applies today. Whether the subject is evolution, climate change, ecology, or any number of other topics, the world does not bend itself to ideology, denial, or wishful thinking. As the science fiction writer Philip K. Dick observed, "Reality is what doesn't go away when you stop believing in it."[15]

Yet one of the great ironies of science is that the more faithfully we attend to reality, the stranger it turns out to be. As the evolutionary biologist J. B. S. Haldane put it, "The universe is not only queerer than we suppose, but queerer than we can suppose."[16]

The history of science reminds us that our willingness to let reality correct us has been gradual and uneven. Resistance to new discoveries has not been limited to religious institutions. Scientific and medical communities themselves have often struggled to abandon established practices even when evidence suggested those practices were ineffective or harmful.

In 1847, the Hungarian physician Ignaz Semmelweis showed that requiring doctors and medical students to disinfect their hands with chlorinated lime dramatically reduced deaths from childbed fever. Many physicians resisted his findings. Likewise, bloodletting remained common for centuries despite little evidence that it was effective and growing evidence that it often caused harm.[17]

Medicine spent centuries relying on theories no better grounded than the classical elements of earth, air, fire, and water. Humorism held that four bodily fluids (blood, phlegm, black bile, and yellow bile) determined health and temperament, making us sanguine, phlegmatic, melancholic, or choleric. Historians of medicine have often observed that until roughly 1900, the average sick person would have been better off avoiding medical care.[18] Controlled experimentation, antiseptic techniques, germ theory, and statistical analysis gradually changed that by tying medical claims to observable results.

The history of medicine reminds us that evidence can overturn long-established assumptions. By learning to correct itself, medicine gave us vaccines, antibiotics, anesthesia, and modern surgery. It transformed human life.

But the history of science is not only a story of correction. It is also a story of discovery, and of the exhilaration that can come when a new explanation suddenly makes the world more intelligible. Weinberg captures that feeling:

> Faced with a puzzling world, people in every culture have sought explanations.... But imagine how Ptolemy must have felt when he realized that he had found a theory of planetary motion that allowed him to predict with fair accuracy where any planet would be found in the sky at any time.... What pleasure Copernicus must then have felt when he was able to explain that the fine-tuning and the looping orbits of Ptolemy's scheme arose simply because we view the solar system from a moving Earth.... How much then the mathematically gifted Kepler must have enjoyed replacing the Copernican mess with motion in ellipses....[19]

Within Christianity, many thinkers responded to such discoveries defensively, asking in various forms what "Athens" (reason, philosophy, and the pursuit of knowledge) could possibly have to do with "Jerusalem" (revelation, faith, and religious tradition).

Liberal religion answers differently. It holds its claims about reality accountable to reason, experience, and evidence. It takes the discoveries of modern science seriously while rereading inherited religious traditions critically—attending to their metaphors, archetypes, and stories—and asking how they can still help us live wisely in the world as we now understand it.

Scientific inquiry is extraordinarily effective at investigating observable, measurable, and repeatable phenomena.

But grief does not hold still for a laboratory. Neither does awe, or love, or the question of how to live.

This is one reason liberal religion continues to draw upon ancient wisdom traditions. They have long wrestled with such questions—not always wisely or accurately, but often with a seriousness that remains worthy of attention.

6

THE SUN DOESN'T RISE

What changes when the whole human story becomes one line?

I remember as a child the first time I really understood that the sun does not rise. The Earth turns. I remember standing outside afterward, looking at the horizon differently. The sun wasn't rising. The Earth was turning.

The next time you watch a sunrise or sunset, try reversing the picture in your mind: instead of imagining the sun moving across the sky, feel yourself being carried with the turning Earth.

A more dramatic shift in perspective came when human beings first traveled beyond Earth orbit and looked back at our planet from deep space. On Christmas Eve of 1968, the crew of Apollo 8 became the first human beings to orbit the Moon. NASA's mission plan had listed photographs of the

Earth under "targets of opportunity," its lowest priority. On their fourth orbit, the onboard recorder captured this exchange:

> Borman: Oh my God! Look at that picture over there! Here's the Earth coming up. Wow, that is pretty.
>
> Anders: Hey don't take that, it's not scheduled.
>
> Borman: (Laughter) You got a color film, Jim?
>
> Anders: Hand me that roll of color quick, will you—
>
> Lovell: Oh man, that's great!
>
> Anders: Hurry, quick—
>
> Lovell: Take several of them! Here, give it to me—
>
> Borman: Calm down, Lovell.

They got the photograph. It became known as *Earthrise.*

We are accustomed to experiencing Earth as the stage on which human history plays out. From lunar orbit, humanity saw our planet as a small, luminous, fragile sphere suspended within the darkness of space. National borders disappeared. Political divisions vanished. The Earth appeared not as an abstraction but as a single living world containing everything we have ever known, loved, feared, built, destroyed, or hoped for. *Earthrise* changed not only how we saw Earth but how we understood our place within its larger story.

The poet Archibald MacLeish captured something of the

emotional and spiritual power of that shift in perspective: "To see the Earth as it truly is, small and blue and beautiful in that eternal silence where it floats, is to see ourselves as riders on the Earth together...."[1]

Fifteen months after the photograph was published, the first Earth Day was celebrated. The photograph functioned almost as a technological form of revelation: a sudden expansion of consciousness made possible by science, exploration, and human imagination.[2]

More than twenty years later, Voyager 1 would offer an even more distant perspective. In 1990 it turned its camera back toward Earth from about 3.7 billion miles away. Reflecting on the resulting image in *Pale Blue Dot*, Carl Sagan wrote:

> Our posturings, our imagined self-importance, the delusion that we have some privileged position in the Universe, are challenged by this point of pale light. Our planet is a lonely speck in the great enveloping cosmic dark.... This distant image of our tiny world underscores our responsibility to deal more kindly with one another, and to preserve and cherish the pale blue dot, the only home we've got.[3]

The interdisciplinary field of Big History asks us to make a similar shift in perspective. It situates human history within the far larger story of cosmic, geological, and biological change, seeing our species as one relatively recent expression of that history rather than as something separate from it.

Thinking at this scale does not come easily. Human beings evolved to think in terms of families, tribes, and seasons—not galaxies and geological epochs. Developing deep time consciousness requires more than information. It requires new habits of imagination and humility, and a moral concern wide enough to match the world we now know we live in.

Long before David Christian coined the term Big History, H. G. Wells saw that modern science was going to require a larger story. Wells is remembered for *The Time Machine* and *The War of the Worlds*. He also foresaw aircraft, tanks, space travel, nuclear weapons, satellite television, and something resembling the World Wide Web.

Wells believed that this expanded perspective was not merely an intellectual achievement but a moral necessity. In the aftermath of the First World War, he argued that lasting peace would remain elusive so long as human beings organized their identities primarily around narrow national, tribal, and competing historical narratives. Humanity needed a story large enough to match the world we had entered.

As Wells wrote:

> There can be no peace now...but a common peace in all the world; no prosperity but a general prosperity. But there can be no common peace and prosperity without common historical ideas.... With nothing but narrow, selfish, and conflicting nationalist traditions...people are bound to drift towards conflict and destruction.[4]

The French Jesuit priest and paleontologist Pierre Teilhard de Chardin reached a strikingly similar conclusion after the Second World War. Writing in 1947, in the shadow of the atomic bombings of Hiroshima and Nagasaki, he argued that humanity had entered a new stage of history in which our survival increasingly depended upon learning to think and act as one planetary community. "The Age of Nations is past. The task before us now, if we would not perish, is to build the Earth."[5]

Wells, Teilhard, and others have called for a story large enough to unite humanity. David Christian's *Origin Story: A Big History of Everything* offers one compelling answer. It locates all of us within a single account running from the birth of the universe to the rise of human civilization. It does not erase cultural differences or local histories; it gives them a wider frame: a shared human context adequate to an interconnected world.

To help make these immense stretches of time more imaginable, Christian encourages us to compress the entire 13.8-billion-year history of the universe by a factor of one billion. Within this thought experiment, events that seem unimaginably ancient suddenly become easier to grasp:

- The Big Bang happened 13 years ago.
- Earth formed 4½ years ago.
- Life emerged almost 4 years ago.
- Dinosaurs disappeared only a few weeks ago.
- *Homo sapiens* appeared about 100 minutes ago.
- Civilization arose a few minutes ago.
- The industrial age began only a few seconds ago.[6]

The whole of recorded human history occupies only the tiniest fraction of the cosmic story.

The late evolutionary evangelist Michael Dowd offers another memorable metaphor for grasping the scale of deep time. Imagine the entire 13.8-billion-year history of the universe compressed into a thirty-volume encyclopedia set, with each page representing one million years.

Within this cosmic library, Earth itself does not appear until Volume 21, and the earliest known life on our planet emerges in Volume 22. Dinosaurs arrive only midway through the thirtieth and final volume before disappearing about 160 pages later.

Modern humans do not appear until the very bottom of the final page: "The entire history of human intelligence, ethics, religious aspiration, and scientific discovery takes up only the last few lines on the last page of the last volume."[7]

Against astonishing odds, a universe that began with hydrogen eventually produced stars, living cells, blue whales, and the music of Bach. A species that arrived on the last line of the last page is now reshaping what happens next for much of life on this planet.

7

STOP BEING ONE PERSON

Who am I when "me" grows larger?

The novelist and essayist David Foster Wallace liked to tell the story of two young fish swimming along when they pass an older fish, who nods and says, "Morning, boys. How's the water?" They swim on for a while until one finally turns to the other and asks, "What the hell is water?"[1]

The point is not that the fish are foolish. It is that the realities that shape us most are often the hardest to see because we have never experienced life outside them. Psychological growth often begins when we become aware of the waters we have been swimming in all along.

Developmental psychologist Robert Kegan offers a helpful way to understand how those hidden waters become visible. Our capacity for reflection grows, he argues, as we learn to turn around and look at what we once saw through. Kegan puts it this way: growth occurs when "the subject of one stage becomes the object of the next."[2]

The *Earthrise* photograph offers another example of what Kegan means. The Apollo 14 astronaut Edgar Mitchell described the shift this way:

> You develop an instant global consciousness, a people orientation, an intense dissatisfaction with the state of the world and a compulsion to do something about it. From out there on the moon, international politics look so petty. You want to grab a politician by the scruff of the neck, and drag him a quarter of a million miles out and say, "Look at that, you son of a bitch."[3]

This phenomenon has come to be known as the "overview effect": a sense of awe and interconnectedness while looking back on Earth from space, sometimes described as a spiritual awakening. Such experiences can be life-changing. But they are not automatic. Traveling into space does not inevitably make people wiser, kinder, or more compassionate. Nor are such experiences confined to astronauts. Moments of expanded perspective are available much closer to home: practicing meditation, standing in a forest, listening to a piece of music that makes everything else go quiet, or holding a sleeping child.[4] Psychedelic experiences can produce the same shift more dramatically—loosening the boundaries of the ordinary self and softening the sense of separation from the world.[5]

However we arrive there, what matters is learning to see

ourselves anew: as part of something larger than the ordinary self.[6]

Understanding interdependence is only the beginning. As our sense of self expands, so can our moral imagination. The more urgent question is what this wider sense of belonging asks of us in practice. In 1962, Rachel Carson's *Silent Spring* helped launch the modern environmental movement, awakening widespread public awareness of humanity's capacity to damage the natural systems upon which life depends. More than six decades later, however, awareness has proved easier to raise than the sustained collective action such knowledge demands.

Part of the difficulty lies in the enormous mismatch between the short time horizons of human attention and planetary timescales. The climate warms slowly enough to be psychologically easy to ignore, but quickly enough to destabilize ecological systems within only a few generations. Most of us rarely notice the atmosphere warming from one day to the next. We notice when a once-in-a-century flood happens twice in a decade, or when wildfire smoke turns the sky orange hundreds of miles from the nearest flames. But even when the consequences become harder to ignore, entrenched economic interests, political polarization, misinformation, and short-term thinking have all delayed meaningful collective action despite overwhelming scientific evidence.

The broad outlines of climate mitigation are already well understood: rapidly transitioning away from fossil fuels, expanding renewable energy and public transportation, designing more walkable and energy-efficient communities, and reducing food waste. No single measure is enough.

Together, these changes point toward lifestyles and social systems more compatible with the ecological realities of a finite planet.

In high-consuming societies such as the United States, living within those limits will take more than better technology. It will take a change in what we value: learning to measure a good life by something other than how much we consume.

Responding to climate change also requires honesty about historical inequality and unequal contributions to the crisis. China is now the world's largest annual emitter of carbon dioxide, while the United States remains responsible for the largest cumulative share of historical CO_2 emissions. Climate change is often discussed as though humanity as a whole contributed equally to the crisis. In reality, the industrialized nations that benefited most from fossil-fuel-driven economic growth have contributed disproportionately to the greenhouse gases now warming the planet.

Yet high living standards do not require equally high emissions. Germany, for example, has Europe's largest economy while producing far less carbon pollution per person than the United States. The challenge is not whether prosperous societies can emit less, but whether we are willing to make the changes required.[7]

As Greta Thunberg notes in *The Climate Book*, many of the communities now most vulnerable to climate disruption have contributed least to creating the problem while often possessing the fewest resources to adapt to its consequences.[8] The journalist and activist Naomi Klein's proposal follows from this: the polluter pays. Fund the transition by requiring that "the people and institutions that have profited most from pollution pay the most to repair the harm it has done."[9] Sounds like a plan to me.

The same pattern of inequality exists within countries as well. The richest 10 percent of the global population produce nearly half of all lifestyle-related carbon emissions, whereas the poorest half produce only about 7 percent. This striking disparity reminds us that responsibility for causing climate change is distributed far less equally than environmental rhetoric often assumes.[10]

Those differences in contribution to the crisis matter. For those of us with high-consumption lifestyles and the resources to change them, reducing our own emissions is part of taking responsibility. That can mean consequential choices such as flying less, driving less or choosing lower-emissions transportation, and reducing the energy our homes use. But the changes we need also go far beyond what any of us can accomplish individually. Responding effectively requires laws, markets, technologies, and public institutions designed to make responsible choices easier and align individual behavior with the common good.

Consider recycling. In my county, you can recycle plastic bottles, jugs, jars, and tubs. You cannot recycle plastic cups, food wrappers, plastic bags, paper towels, or takeout containers. Many people put them in anyway. This does not get them recycled; it makes the whole process more expensive and slower for everything else.

Nationally, "only 8 out of 100 plastic items we put in the recycling bin actually get recycled. This means that the other 92 plastic items...are burned or end up in landfills...."[11] Rather than placing the burden almost entirely on individual consumers, we need economic and political structures that make products easier to repair, reuse, and recycle while making sustainable living easier, more affordable, and more widely accessible. Living sustainably should become an ordinary part of everyday life, not an exceptional burden placed on individuals.

That also means recognizing that industries benefiting from the status quo may resist changes that shift more responsibility from consumers back to producers, especially when profits depend on frequent replacement, short product lifespans, or hard-to-repair products. A more sustainable economy has to address the whole life of a product, from how it is designed and made to whether it can be kept in use and out of the landfill.

Let me offer one small but revealing example. When my city began offering free composting a few years ago, my family signed up immediately. We had talked about composting for years, but like many people, we never quite got around to it until the program made it easy. Now, every Tuesday morning, I simply set our compost bin at the curb. Each week the composting service empties it and returns the bin freshly cleaned and ready for the week ahead. Each spring, residents are also offered free compost made from what we all put out at the curb.

That experience reinforced an important lesson. Sustainable living becomes far more achievable when our institutions make responsible choices easy, convenient, and affordable. An individual focus asks, "What can I do?" Ecological consciousness asks, "What kinds of communities and systems make it easier for all of us to do the right thing?" The goal is not merely to encourage better individual behavior, but to design systems that make the better choice the easier choice, and make environmentally harmful choices more difficult and costly.

I've seen the same principle at work in my own congregation. We are installing solar panels not only because they reduce emissions, but because they save money that can be redirected toward our mission. Most of all, it means our building increasingly runs on the very thing we say we believe in.

Although many of us wish the movement for climate justice were much further along than it is, our predicament would be far worse without the environmental movement of the past half century. We remain on track to exceed 1.5°C of sustained global warming above pre-industrial levels, a threshold associated with increasingly severe climate impacts. Even so, as Rebecca Solnit and Thelma Young Lutunatabua observe in *Not Too Late: Changing the Climate Story from Despair to Possibility*, had our collective future been left to cynical and bad-faith actors, we would now be hurtling toward 4°C or more of warming—plunging into a climate catastrophe far worse than the one we already face.[12]

Progress has been painfully slow, but it has been real. Every species protected, every ecosystem restored, every ton of carbon kept out of the air, and every community better prepared for what's coming reduces suffering and leaves the future more open than we found it. The challenge is not to choose between realism and hope, but to practice what Rebecca Solnit has called "hope in the dark": acting responsibly not because success is guaranteed, but because the future remains unfinished.

One of the most important ways we practice that kind of hope is by refusing to face these challenges alone. The best advice about climate justice I have ever received is this: if you feel overwhelmed by what you can do as one person in the face of such an enormous need for systemic change, "Stop being one person."[13] Join a group or movement larger than yourself and work for systemic change.

Find the group that fits you—and your place within it. Not everyone needs to march or speak in public; movements also depend on the quieter work of keeping things running behind the scenes. Does a nearby congregation have a climate justice group? The Sunrise Movement is youth-led. Citizens' Climate Lobby works through nonpartisan advocacy with elected officials. Extinction Rebellion practices nonviolent civil disobedience.

History often changes when people stop being one person.

8

CHOOSING EARTH

What would it mean to stay?

Imagine spending six months aboard the International Space Station, circling Earth every ninety minutes and watching sixteen sunrises and sunsets each day. That was the experience of British astronaut Tim Peake.

When he returned home, he wrote *Ask an Astronaut: My Guide to Life in Space*. In it, he offers a bold prediction: "It is no longer a question of if we will colonize the Moon and Mars, but when."[1] Many people (including some of the wealthiest and most influential figures in the world today) share Peake's confidence.

I hope they are right. Exploring the solar system and eventually establishing permanent human settlements beyond Earth would be among humanity's greatest achievements. Over the long arc of history, we may indeed become a multi-planetary species.

But for the foreseeable future, our fate remains inseparably tied to a single planet. Even under the most optimistic scenarios, colonies on the Moon or Mars would remain tiny, fragile, technologically dependent outposts while billions of human beings continue to live on Earth and depend upon its ecological systems. Dreams of becoming a multi-planetary species should therefore inspire exploration, not distract us from the urgent work of caring for the only world that can sustain human civilization for generations to come.[2]

Don't get me wrong: I'm pro-space. I continue to be awestruck by the images from the James Webb Space Telescope. And I've been inspired by the Artemis missions, which are carrying human beings farther from Earth than we've traveled in more than half a century.

But alongside that wonder, we also need to face a harder truth: Space is hostile to human life. One of the best guides to that world is Kelly and Zach Weinersmith's *A City on Mars: Can We Settle Space, Should We Settle Space, and Have We Really Thought This Through?* Their conclusion is refreshingly blunt: space is terrible.

The problem is that the images we see (the stark beauty of the Moon or the sweeping vistas of Mars) can be deceptive. They resemble deserts: harsh, perhaps, but familiar. In actuality, they are nothing of the sort.

The Moon's surface is covered not in soft sand but in razor-sharp, electrostatically charged dust that clings to everything, abrades machinery, and can damage human lungs. Mars is even less forgiving. The average temperature hovers around -80°F. There is no breathable atmosphere.

The soil is laced with toxic chemicals, and massive dust storms can block out the sun for weeks at a time.

By comparison, even Antarctica (one of the harshest environments on Earth) is vastly more hospitable to human life than anywhere else in the solar system. As the Weinersmiths observe, choosing Mars over even a severely climate-changed Earth would be less like moving somewhere harsher than like leaving a messy room to live in a toxic waste dump.[3]

That comparison reminds us how easy it is to take our own planet for granted. Earth is an astonishingly rare oasis: a world with breathable air, liquid water, a protective atmosphere, fertile soil, and a biosphere capable of sustaining billions of human beings alongside millions of other species—the very conditions under which our own species evolved. Before dreaming of becoming a multi-planetary civilization, we would do well to care for the extraordinary home we already have.

As the parent of a preschooler, I already have plenty to worry about here on Earth. Now imagine trying to raise a child in space. Instead of wondering what they're eating, how much screen time they're getting, or whether they're getting enough sleep, you would be confronting something far more fundamental: whether the environment itself can support healthy human development.

In low gravity, bones lose strength month after month, even with hours of daily exercise. Add higher levels of radiation, elevated carbon dioxide, and an environment utterly unlike anything our bodies evolved to inhabit, and the real question becomes whether human beings could live there safely across generations.[4]

We often imagine space colonization as an engineering problem. In reality, it is also a biological one. Human beings did not evolve for life on the Moon or Mars. We evolved on Earth. Our bones, muscles, lungs, immune systems, and circadian rhythms are all exquisitely adapted to this planet, as are the conditions under which children develop. Before we imagine abandoning Earth, we should remember that our bodies were shaped here.

Even under the bleakest climate scenarios (rising seas swallowing coastal cities, intensifying heat rendering entire regions difficult to inhabit, ecosystems unraveling, crops failing, and millions of people displaced) Earth would remain incomparably more hospitable to human life than anywhere else in our solar system.

Mars is not a backup plan. At best, it is a centuries-long, high-risk experiment. A deeply wounded Earth would still be the only world in this solar system where you could step outside without a spacesuit and remain alive. Re-creating even a fraction of those conditions on another world would require staggering amounts of time, energy, and technological intervention.[5]

Choosing Earth means recognizing reality. This planet is our home and the only known world capable of sustaining human civilization and the astonishing diversity of life we share.

Why does all this matter? As Adam Becker argues in *More Everything Forever: AI Overlords, Space Empires, and Silicon Valley's Crusade to Control the Fate of Humanity*, a handful of extraordinarily wealthy and influential technology leaders have made expansion into space central to their visions of humanity's long-term future. Humanity may indeed explore and eventually settle beyond Earth. The deeper question is what we expect that expansion to solve. If we carry the same habits, inequalities, and unsustainable ways of living with us, another world will not save us from ourselves. Becoming Earthlings is not a choice between Earth and space. It is the work we will have to carry with us wherever we go.[6]

Some of the world's most powerful technology leaders have been remarkably open about their ambitions for humanity's future in space. Jeff Bezos, the founder of Amazon and the space company Blue Origin, has imagined a tomorrow in which a trillion human beings live beyond Earth. He argues that only by expanding into space can humanity continue to grow and avoid stagnation here at home.

Elon Musk, the CEO of SpaceX and Tesla, has likewise spent years arguing that humanity must become a multi-planetary species (especially by settling Mars) if we hope to safeguard our long-term survival against catastrophes on Earth, whether environmental collapse, nuclear war, a rogue asteroid, artificial intelligence gone awry, or some other existential threat we cannot imagine.

Bezos, for his part, offers a different account. Where Musk focuses on escaping potential catastrophe, Bezos presents space as a way to preserve Earth. His concern is that continued population growth and rising energy demands will eventually place unsustainable strain on our planet. His proposed solution is not primarily to settle Mars but to relocate much of human industry off Earth: building infrastructure on the Moon, mining extraterrestrial resources, and constructing vast space habitats.

In this vision, Earth remains precious precisely because it is no longer the primary site of human civilization. It becomes a protected garden world (a place to preserve, cherish, and restore) while much of humanity's industry, manufacturing, and perhaps eventually even population expands into space.[7]

Bezos and Musk tell different stories, but they share an important underlying assumption: that humanity's long-term thriving depends primarily on expanding beyond Earth. Here, however, is the harder truth. Whether we are talking about a Moon base, a space station, or a colony on Mars, history gives us every reason to believe we will carry our deepest problems with us. As the saying goes, "Wherever you go, there you are."[8] We do not magically leave behind our habits, our egos, our systems of inequality, and our tendencies to compete over scarce resources because we have crossed into orbit. There is no technological shortcut around the moral, political, and spiritual work of becoming a more just, compassionate, and sustainable civilization.

Space may offer new frontiers, but it cannot offer a new human nature. Building a new world would mean slowly recreating the fragile social systems we developed here (trust, accountability, shared norms) while still contending with the human tendencies that made those systems necessary in the first place. The technologies required to reach another world may be extraordinary. The greater challenge will remain the one humanity has always faced: learning to live wisely with one another.[9]

The visions offered by Bezos, Musk, and other space entrepreneurs also deserve to be evaluated in light of the ethical records of the institutions they have built here on

Earth. For instance, federal workplace-safety investigators found that Amazon warehouse workers faced a high risk of serious musculoskeletal injuries from frequent lifting, heavy loads, awkward movements, and long hours—conditions that the Occupational Safety and Health Administration (OSHA) concluded were "designed for speed but not safety."[10] As Adam Becker puts it, Bezos is all but saying: Don't look behind the curtain—"at the horrifying labor conditions at the local Amazon fulfillment center. Look at the shiny rocket.... Everything will be better in space."[11]

Elon Musk's Department of Government Efficiency dismantled much of USAID's humanitarian infrastructure, terminating programs that provided food, vaccinations, HIV and tuberculosis treatment, maternal and child health care, and other basic services around the world. By July 2026, former USAID global-health chief Atul Gawande estimated that the cuts had contributed to roughly 700,000 deaths and deprived some 95 million people of basic health-care access.[12] We are now asked to trust Musk to build a safe future for us on Mars. If today's technological elites cannot exercise power wisely here on Earth, why should we assume they will govern more wisely there?

Such themes find vivid expression in two television series: *For All Mankind* and *The Expanse*. *For All Mankind* imagines an alternative history in which the space race never ended. What emerges is not a utopia, but a new frontier shaped by geopolitical rivalry, labor struggles, and growing tensions between Earth and its off-world settlements on the Moon and Mars. *The Expanse*, set centuries ahead, depicts a fully colonized solar system where familiar patterns of power persist: Earth and Mars compete for dominance, while workers in the asteroid belt are exploited for the resources that sustain civilization. In both stories, technology advances, but human nature—and the political, economic, and social systems we create—remain strikingly familiar.

We dream about terraforming Mars. What if we devoted that same imagination, determination, and technological ingenuity to preserving the habitability of Earth? If we truly believe we can one day transform a frozen, irradiated desert into a place fit for human life, surely we can protect the planet that already is.

We already know how to reduce emissions, expand renewable energy, protect biodiversity, restore ecosystems, and build more sustainable communities. The challenge is to act at the scale the crisis requires.

Choosing Earth means more than remaining on one particular planet. It means becoming the kind of civilization capable of caring for the extraordinary world that has already given us everything. That choice is being made now. Scientists continue to warn that "every key climate indicator is flashing red."[13] Yet governments and industries around the world continue to expand fossil fuel production and delay changes that the evidence increasingly demands.[14]

This reflects more than a disagreement about economic policy. One vision treats the Earth's living systems primarily as resources to be extracted for short-term economic gain. The other recognizes that human flourishing ultimately depends upon the long-term health of the ecological systems that sustain all life. Choosing Earth calls us toward the latter.

Politics, however, is only part of what is happening. Even

as some governments double down on fossil fuels, renewable energy is advancing rapidly around the world—a genuine reason for hope.

One of the clearest expressions of that hope appears in Bill McKibben's *Here Comes the Sun: A Last Chance for the Climate and a Fresh Chance for Civilization*. He argues that the change we need is already underway. For the first time in human history, we are discovering how to power our civilization not by burning the Earth's ancient past, but by turning once again to the energy that has sustained life here all along: the sun.

Renewable energy is now an economic reality. In many parts of the world, solar and wind power have become the least expensive sources of new electricity, and they continue to expand rapidly. The technological path out of the climate crisis is available now. The question is no longer whether we can make this transition. It is whether we will.[15]

McKibben is far from alone in proclaiming this good news. *Science*, one of the world's most respected scientific journals, named the "seemingly unstoppable growth of renewable energy" its Breakthrough of the Year, recognizing that the transition to clean energy is no longer a distant hope but an accelerating global reality.[16]

Revolutions often look impossible—until suddenly they look inevitable. For years, solar energy seemed like a promising but marginal technology, advancing only incrementally. Then the curve changed. Growth became exponential. Today, solar is expanding at a pace few experts predicted even a decade ago, transforming what once seemed like a distant alternative into one of the dominant forces reshaping the global energy system.

Much of this growth, however, is happening outside the United States. In 2024 alone, China installed more solar and wind capacity than the rest of the world combined. More broadly, much of the global progress celebrated by *Science* as the 2025 "Breakthrough of the Year" is being driven substantially by the rapid expansion of renewable energy in China and India. For the United States, this global transition should serve as a wake-up call: we can help lead it or continue falling behind.[17]

What would it look like for the United States to help lead? One answer is a Green New Deal.[18] As economists such as Paul Krugman have argued, it would be not only an environmental undertaking but also a project of security and peace.[19]

Because fossil fuels are concentrated in particular places, access to them can be controlled, disrupted, and fought over. Consider the Strait of Hormuz, one of the world's most important oil-shipping chokepoints. Sunlight and wind are different. Although the technologies that harness them require global supply chains, the energy itself is widely available rather than concentrated in a handful of places. The more we rely on energy from the sun and wind, the less vulnerable we become to distant conflicts, volatile fossil fuel markets, and the geopolitical leverage of petrostates. As Bill McKibben memorably observes, "Sunlight travels 93 million miles to reach the earth—none of them through the Strait of Hormuz."[20]

But the transition will require more than technology and policy. Choosing Earth is a shared practice, and liberal religion has a role here beyond public advocacy: helping form people who pay attention to the living world, act alongside others, and keep going when the work is hard. A promising tomorrow will be built by people who choose, together, to care for the home that made us.

So even amid frightening and heartbreaking events, may we be among those who stay, show up, and refuse to give up on this fragile, miraculous home we call Earth.

PART II

MEETING THE MORE-THAN-HUMAN WORLD

Part I explored what it means to become Earthlings. Part II turns to the more-than-human world with which we share the Earth.

Modern life often encourages us to see the more-than-human world as little more than background scenery or raw material for human purposes. Forests become lumber. Rivers become resources. Animals become commodities. Plants become landscaping. The more-than-human world fades into the backdrop of our lives, noticed only when it proves useful—or inconvenient.

Yet the closer we look, the more that familiar picture dissolves. Over recent decades, discoveries across the life sciences have revealed a world alive with astonishing complexity. Forests are sustained by relationships among trees, fungi, microbes, and animals. Plants perceive and respond to their surroundings in ways once thought impossible. Every organism inhabits its own perceptual world, and those worlds continually meet and overlap.

Science offers one way of learning to pay attention. Many Indigenous traditions have long understood the Earth as a web of relationships rather than a collection of objects. Science and Indigenous knowledge can arrive at similar insights by different roads.

None of this means that trees think exactly like humans or that plants possess human consciousness. It does suggest that life on Earth is bound together more tightly, and in stranger ways, than earlier generations imagined.

Meeting the more-than-human world requires both critical thought and closer attention. Wonder can be one of science's companions. The more closely we attend to the living world, the harder it becomes to imagine ourselves as separate from it. That is where belonging begins.

9

HIDDEN IN PLAIN SIGHT

How much life have I mistaken for stillness?

While watching a recording of *Springsteen on Broadway*, I wasn't surprised to hear that one of Bruce Springsteen's most vivid childhood memories centered on a tree: a magnificent copper beech in the front yard of his childhood home, a place where he played, climbed, dreamed, and found refuge from a world he already sensed was complicated. Listening to him, I was reminded how often our memories are rooted in trees.

Let your own mind drift back for a moment. Can you picture a tree that has been significant in your life? Can you remember what it looked like in different seasons? Perhaps you can still feel its bark beneath your hand—or the scent of its leaves or blossoms in the spring.

I can still picture the dogwood tree in the front yard of

my childhood home, the magnolia tree at my maternal grandmother's house, the pecan trees in my paternal grandparents' yard, and the towering oak at my elementary school, its branches draped with long curls of Spanish moss.

But you don't have to grow up with a tree for it to become a beloved companion. Although I had lived in downtown Frederick for only a few years, I was surprised by the grief I felt when the massive weeping willow near the Baker Park Bell Tower fell during a storm. It felt fitting that neighbors quickly gathered for both a vigil and a funeral to honor a tree that had been part of so many lives for more than sixty years. Does mourning a tree seem strange? The more I learn about trees, the more natural—and appropriate—that response seems.

And as significant as a single tree can be, an even deeper kind of magic emerges when trees have the time and space to grow together over centuries, even millennia. I can just as easily recall beloved forests: the acres of woods behind one of my childhood homes in South Carolina, the forests surrounding my childhood summer camp in Western North Carolina, and the holy majesty of Muir Woods, where I often wandered among ancient redwoods while completing my training as a spiritual director in the San Francisco Bay Area.

That affection for trees also has deep roots in the Unitarian tradition. Henry David Thoreau's love of trees and forests was legendary. Even the subtitle of his most famous book points in that direction: *Walden; or, Life in the Woods.* There is also something fitting about Thoreau's decision to remain rooted in his beloved hometown of Concord, Massachusetts. Like a tree, he drew stability, depth, and richness from staying put.

As Richard Higgins writes in *Thoreau and the Language of Trees*, "Thoreau wrote prolifically about trees for a quarter century, from 1836 to 1861. He observed them closely, knew them well, and described them in detail.... He saw them as his friends and even as 'distant relations.'"[1] Thoreau himself once exclaimed, "What cousin of mine is the shrub oak!"[2] For Thoreau, trees were never merely scenery. They were family.

Spend a few moments with these passages from Thoreau's writing about trees, and notice whether any of them match something you've seen yourself:

- In winter, Thoreau marveled at the remarkable symmetry of a hemlock.
- Spring filled him with wonder at the sudden freshness of the world as new leaves appeared, as though "Nature spreads her umbrellas!"
- Summer delighted him with the fragrance of meadows and forests and the fiery glow of sunlight shining through the cones of a black spruce.
- Perhaps autumn taught him the deepest lesson. Watching leaves fall, he reflected on how trees surrender themselves: "How poetically, how like saints or innocent and beneficent beings, they give up the ghost!"[3]

A spiritual practice is just this kind of attention: to watch one tree until it stops being scenery and becomes itself.

One remark captures Thoreau's love of trees especially well. After Harvard College appointed a new professor of natural history, Thoreau thought it made no sense to lecture on oaks while the best specimens in Massachusetts were being cut down: "It is like teaching children Latin and Greek, while we burn the books printed in those languages."[4]

Thoreau's words returned to me near the end of *Springsteen on Broadway*, when Springsteen returns to his childhood hometown and discovers that the beloved copper beech tree of his youth—the tree that had stood long before he was born and that he had always imagined would outlive him—had been cut down. He describes the loss as unexpectedly devastating: a reminder that even the landmarks we assume will anchor our lives can disappear. Some trees become landmarks not only in the landscape but in our lives.

To delve more deeply into the mysteries of trees, let us turn from the nineteenth-century forests of Concord and the twentieth-century suburbs of New Jersey to the present day. In recent years, no one has taught me more about trees than the forester Peter Wohlleben. His book *The Hidden Life of Trees* is one of those rare books that can shift how you experience the world.

One of the first keys to understanding the hidden life of trees is recognizing that, compared to humans, trees live on an entirely different timescale. One of the oldest known trees on Earth, a Norway spruce in Sweden, has been estimated to be more than 9,500 years old. With such extraordinary longevity, trees can afford a leisurely pace. As Wohlleben notes, trees experience time on a scale almost unimaginable to us. Their electrical impulses, for example, move through their roots at only about one-third of an inch per second.[5] To us, that seems impossibly slow. To a tree that may live for thousands of years, it is simply how life moves.

The slow pace of trees can lead us to mistake them for passive, almost inanimate beings. But those slow-moving electrical impulses offer one of our first glimpses into the hidden life of trees. Trees communicate in surprising ways. They send electrical signals, release airborne chemical messages, and even respond to the tastes and smells of the world around them. When a giraffe begins feeding on an African acacia, the tree releases airborne chemicals that warn neighboring trees of the threat. Those nearby trees respond by producing defensive compounds before the giraffe reaches them. Trees can also "taste" the saliva of leaf-eating insects and respond by releasing chemicals that attract predators of those insects.[6] From the outside, trees may look silent and still. But they have a lot going on.

More astonishing still are the intricate—even intimate—relationships among trees growing together in natural forests. Wohlleben calls this underground network the "wood wide web."[7] Through vast fungal networks that connect their roots, trees share nutrients, exchange chemical signals, and, in some cases, even sustain one another. One of the most remarkable discoveries is that living trees have been found nourishing the stumps of long-felled neighbors, sending them sugars and other nutrients through these underground connections. Wohlleben wonders whether some of those ancient stumps might even be the parents of the trees that now surround them.[8]

There's more. Wohlleben points to evidence that the roots of trees are far more than simple anchors. They may also serve as something like an information-processing center. If trees can learn from experience, they must also have some way of storing information. Scientists do not know exactly how this works, but researchers at the University of Bonn have shown that when roots encounter obstacles such as toxic chemicals, rocks, or waterlogged soil, they process that information, alter their growth, and grow around the hazard.[9]

The findings become even more surprising. The plant biologist Monica Gagliano has run experiments on the sensitive mimosa—not the champagne-and-orange-juice cocktail, nor the ornamental mimosa tree, but a tropical creeping herb. In one experiment, mimosa plants were repeatedly dropped a short distance. At first they folded their leaves defensively, but after enough harmless drops they stopped. When tested weeks later, they still did not fold their leaves in response to the drop—suggesting they had retained what they learned.[10]

None of this proves that trees possess consciousness in the way humans do. It does, however, encourage us to ask whether consciousness exists in different degrees and forms rather than as an all-or-nothing phenomenon. Human self-awareness need not be the standard against which every other kind of mind is measured. A dog, an octopus, a tree, and a single cell respond to their surroundings in radically different ways. In some cases, we have strong reason to believe there is subjective experience; in others, we do not yet know whether responsiveness amounts to awareness. The point is not to declare everything conscious, but to resist drawing the boundary too quickly. Perhaps the Earth is not neatly divided into beings with minds and beings without them, but contains a far wider range of sensing, memory, communication, and response than we have learned to recognize.

At this point, some readers may wonder whether all of this is nothing more than anthropomorphism: projecting human traits onto trees and plants. It is fair to ask. But the idea that plants possess something like perception, memory, or even a primitive form of intelligence is hardly new. Less than two years before his death in 1882, Darwin published *The Power of Movement in Plants*, arguing that the tip of a plant's root "acts like the brain of one of the lower animals... receiving impressions from the sense organs and directing the several movements."[11]

When we take Darwin's theory of common descent seriously, the remarkable continuities we find across the living world should not surprise us. All known cellular life shares a common evolutionary ancestry. In evolutionary terms, Thoreau's language of kinship turns out to be more than metaphorical: humans and trees retain many homologous genes inherited from ancient common ancestors.[12] We can therefore expect at least some of the capacities we associate with animals to have simpler or differently expressed analogues elsewhere in the tree of life. From that perspective, it seems less strange that plants might possess their own forms of memory, communication, sensation, and learning. Plants and animals remain distinct branches of the tree of life, but some capacities we once associated primarily with animals appear in very different forms elsewhere in the living world.

The hidden life of trees is far richer and more complex than most of us were ever taught. Beneath the soil lies a world of communication, cooperation, memory, and relationship that we are only beginning to understand. Wohlleben's greatest gift may not be the facts at all. It's that he teaches us to pay attention.

Wohlleben closes with an invitation that has stayed with me:

> I encourage you to look around where you live. What dramas are being played out in the wooded areas you can explore? How are commerce and survival balanced in the forests and woodlands you know? ... Slow down, breathe deep, and look around. What can you hear? What do you see? How do you feel?[13]

I took that invitation to heart. Around the same time I was reading Wohlleben, I learned that less than a tenth of a mile from my house stood the largest sugarberry tree in the state of Maryland.[14] Seventy-seven feet tall with a trunk circumference of eleven feet, it had quietly stood there all along while I walked past it countless times without realizing what stood before me. Once I learned its story, I could never see it the same way again.

Wherever trees grow, there are some worth knowing. If you were to visit Frederick, Maryland, I would point you to the great English elm at Frederick High School and the magnificent ginkgo on West Second Street, one of the oldest and largest in the United States. Remarkable trees often surround us, hidden in plain sight.[15]

Remarkable as these champion trees are, one of the central lessons of *The Hidden Life of Trees* is that the deepest wonder lies not in solitary giants but in ancient forests, where lives have become intertwined over centuries. To protect old-growth forests is to protect one of Earth's oldest and most extraordinary forms of community.

To learn from trees is also to learn to think in centuries rather than seasons. The Kentucky farmer, poet, and environmental prophet Wendell Berry has spent a lifetime calling us to think on longer timescales. In his poem "Manifesto: The Mad Farmer Liberation Front," he invites readers to imagine lives shaped by responsibilities that extend far beyond their own lifetimes. Throughout his work, he challenges us to measure success not only by what we can harvest for ourselves, but also by what we leave to generations we will never meet.

To plant trees we will not live to see mature, or protect forests whose greatest gifts will go to people we will never meet, is one way to become a good ancestor.

10

RELATIONSHIPS ALL THE WAY DOWN

Where do I actually end?

A few years ago, *The New York Times* published a major article titled "The Social Life of Forests." Its subtitle captured its central insight: "Trees appear to communicate and cooperate through subterranean networks of fungi." That final phrase deserves our attention. Much of what happens within a forest (the exchange of nutrients, chemical signals, and other forms of communication) is made possible by fungi, whose vast underground networks connect the roots of plants. Largely hidden beneath our feet, fungi are among the indispensable architects of forest life.[1]

What drew me more deeply into this hidden world was the book *Entangled Life: How Fungi Make Our Worlds, Change Our Minds & Shape Our Futures* by the mycologist Merlin

Sheldrake. His book turns our attention beneath the forest floor, where vast fungal networks quietly connect, nourish, and sustain much of the life above them. What appears at first to be a forest of separate trees is really a web of living relationships.

Sheldrake is a quirky young scientist, and I suspect there's much more to come from him. To give you a sense of it: he celebrated the publication of *Entangled Life* by cooking and eating oyster mushrooms that had grown out of a copy of his own book.[2] That's some thoroughgoing commitment to your work.

The more I learn—from biology and ecology, quantum physics, and my own meditation practice—the clearer it becomes that interdependence is not merely an ethical aspiration. It is a description of the world. We are bound up with one another, with the countless other forms of life on this planet, and with the ecosystems that keep us alive. Our task is not to create that connection but to wake up to it.

As I write these words in the summer of 2026, smoke from massive boreal forest wildfires in Canada has once again drifted hundreds of miles south, dimming the sun and turning the sky a hazy gray here in Maryland. It is a vivid reminder that forests are not distant places existing somewhere "out there." Their health reaches into our lungs, our communities, and our daily lives. The hidden relationships that sustain forests also bind our lives to theirs.

Our growing understanding of the hidden partnerships within forests is only the latest confirmation of John Muir's observation: "When we try to pick out anything by itself, we find it hitched to everything else in the universe."[3] Living from that felt sense of interdependence can change how you move through the world.

Take one slow breath. Air from the world enters you and returns to it. Where, exactly, do you end?

For now, the part of nature we are picking up—and finding hitched to everything else—is the kingdom of fungi. And we have only begun to discover how extraordinary that hidden world really is. It includes both mushrooms (the familiar fruiting bodies that emerge above ground) and mycelium, the vast underground networks of branching threads that connect, communicate, and sustain forest life.

One indication of how much remains to be discovered comes from an exchange between the primatologist Jane Goodall and the mycologist Paul Stamets. After learning what fungi might be capable of, she encouraged him with these words: "Don't stop. You're with fungi where I was with the chimps when I started."[4]

As Sheldrake explains, scientists estimate that Earth is home to between 2.2 and 3.8 million fungal species (roughly six to ten times as many as there are plant species), but only about six percent have been formally described.[5] In other words, more than ninety percent of Earth's fungal species remain undocumented by science.

Even so, we already know enough to recognize that fungi are indispensable to life on Earth. More than ninety percent of plants depend on mycorrhizal fungi, which form intimate underground partnerships with plant roots. As Sheldrake observes, "The more we learn about fungi, the less makes sense without them."[6]

Indeed, fungi are so common throughout the plant kingdom that they are "a more fundamental part of plants than fruit, flowers, leaves, wood, or even roots." In this symbiotic partnership, plants feed fungi with the sugars they produce through photosynthesis. In return, the fungi scour the soil far more effectively than plant roots can, gathering water and nutrients and delivering them back to their plant partners. These same fungal networks often connect neighboring plants, allowing nutrients, water, and even chemical signals to pass through the mycelial web.[7]

The relationship is so intimate that it becomes increasingly misleading to think of plants as simply having roots. The word *mycorrhiza* itself means "fungus-root." That single word points to a larger truth: the living world is best understood not as a collection of separate individuals that occasionally interact, but as relationships all the way down.[8]

From a human perspective, it is understandable that mushrooms are often the first thing that comes to mind when we think of fungi. They are visible, above ground, and undeniably fascinating.[9] I especially love that the Potawatomi language includes the word *puhpowee*, which means "the force that causes mushrooms to push up from the earth overnight."[10] I first encountered this beautiful word in Robin Wall Kimmerer's *Braiding Sweetgrass*, and I love that a language would have a single word for such a mysterious and enchanting phenomenon.

We are only scratching the surface of the entangled lives of fungi. Fungi have already transformed medicine, giving us penicillin and compounds such as cyclosporine, which helped make modern organ transplantation possible. As scientists continue studying fungi, they are uncovering new possibilities for medicine, environmental repair, and fields we have not yet imagined. Some of the most important biological discoveries of the coming decades may well come from fungi.

Fungi have captured public attention for another reason as well: certain species can temporarily reshape the networks within the human brain. Imaging studies suggest that under the influence of psilocybin (the psychoactive compound found in so-called "magic mushrooms"), brain regions that do not ordinarily communicate begin interacting in striking new ways. The result is a temporary increase in connectivity, which may help explain why many people report fresh insights, enhanced creativity, a diminished sense of separation from the world around them, and lasting benefits for conditions such as depression, anxiety, and addiction.[11]

One final discovery is especially remarkable. Although fungi do not have brains in the way we usually understand them, a growing body of research shows that their mycelial networks actively sense and respond to their surroundings. The long branching filaments called hyphae are anything but passive. As Sheldrake puts it, they interpret the chemical signals of others "to negotiate a series of complex trading relationships with trees; to knead away at stores of nutrients in the soil; to have sex; to hunt; or to fend off attackers."[12] Even more remarkably, experiments involving mazes and blocks of wood suggest that mycelium may possess a kind of directional memory, although scientists are still working to understand how that memory functions.[13]

The implication is fascinating: "Intelligent behaviors can arise without brains. A dynamic and responsive network is all that's needed."[14] If so, then perhaps intelligence, learning, and responsiveness are more widely distributed throughout the more-than-human world than we have long assumed. The more we learn, the more we realize how much remains beyond our understanding. As with the hidden life of trees, the entangled life of fungi is far richer and more complex than our ordinary human perspective allows us to see.

Although we still have so much to learn about fungi, I find myself agreeing with Sheldrake's larger point. His study of fungi has led him to rethink some of our most fundamental assumptions about "evolution, ecosystems, individuality, intelligence, and life."[15] The same has been true for me: the more closely I look, the less separate everything appears.

Relationships are not something living things enter into; they are what life is made of. And those relationships include the nonliving world. Life draws on air and water, minerals and sunlight, and the remains of lives that came before, though different organisms depend on different sources. Wisława Szymborska approaches this interconnected world without our usual labels in her poem "View with a Grain of Sand." The grain does not know itself as a grain, and the lake does not see itself as a lake. Those names —and many of the boundaries they mark—belong to us more than to the world itself.

That may be the lesson running through both the hidden life of trees and the entangled life of fungi: the relationships that sustain life often cross the boundaries our categories draw.

11

SAME ROOM, DIFFERENT WORLDS

What is real that I'll never sense?

One mindfulness meditation practice is called Six Sense Noting.[1] This practice begins with the five senses many of us learned about in school (sight, hearing, smell, taste, and touch) and then adds a sixth in the traditional Buddhist framework: the mind itself, through which we experience thoughts and other mental phenomena.

The practice is simple. Notice whatever stands out most in your awareness at that moment and mentally label it according to one of six categories:

- Seeing
- Hearing
- Smelling

- Tasting
- Touching
- Thinking

Then continue, moment by moment, noting whatever is most prominent in your awareness as it changes.

In Buddhism, these correspond to what are often called the Six Sense Gates: eye, ear, nose, tongue, body, and mind. Within this framework, every moment of human experience enters awareness through one or more of them.

This traditional Buddhist map is not meant to be a modern scientific inventory of the senses. Contemporary science distinguishes other forms of sensory experience as well, including balance and proprioception (our awareness of the position and movement of our own bodies).[2] Even this expanded list remains centered on human experience.

Other animals perceive aspects of the world that lie entirely beyond us, through senses that do not fit neatly within our categories. Our senses reveal a vast world—but not the whole of reality. Each species inhabits its own perceptual world, shaped by the capacities evolution gave it.

Few people have done more to help me see past the limits of my own senses than the Pulitzer Prize-winning science writer Ed Yong in his book *An Immense World: How Animal Senses Reveal the Hidden Realms Around Us*. I found it riveting.

He begins with a simple thought experiment. Imagine that you are the only human being in a large room, sharing it with a menagerie of other creatures. There's an elephant standing at one end of the room, a mouse scurrying beneath the furniture, a robin perched on a windowsill, an owl resting on one of the overhead beams, a bat hanging upside down from another beam, a rattlesnake slithering across the floor, a spider lurking in its web in a high corner, a mosquito buzzing nearby, and a bee.

It's an improbable scene, but stay with me. The point is to imagine how each of these creatures experiences the very same room.

Let's start with the mosquito, which would be drawn toward the carbon dioxide in your exhalations and the scent of your skin, with your body heat helping guide it as it came closer. When it lands on your arm, the sound of you slapping it away would most likely disturb the mouse.

The mouse's squeak of alarm would be too high-pitched for the elephant to hear, but easily audible to the bat. If the elephant answered with a deep rumble, it would be too low for either the mouse or the bat to hear, but the snake would feel its vibrations through the floor. We humans could perceive neither the mouse's ultrasonic squeak nor the elephant's infrasonic rumble (below the range of human hearing), but we would hear the robin's song, even if we could not appreciate its full complexity.

The robin's breast appears red to us, but not to the elephant, whose color vision is limited largely to blues and yellows. Meanwhile, the bee would be drawn to ultraviolet patterns on flowers outside the window: patterns entirely invisible to human eyes.

The spider, meanwhile, would remain oblivious to most of this sensory commotion unless the mosquito happened to vibrate its web, triggering it to strike. But the spider would remain equally unaware of the bat's sonar until the bat swooped down and caught it with remarkable precision.

If the days were growing shorter and colder, the robin would also sense subtle shifts in the Earth's magnetic field, telling it that the long journey south had begun.[3]

They all inhabit the same room. But each lives in a different world. These examples offer only a small glimpse of the extraordinary diversity of perception in the living world.

The scientific term for this phenomenon is *Umwelt*, a German word meaning "environment." In contemporary biology, it refers to the perceptual world available to an organism through its particular senses. Each creature notices some features of reality and remains oblivious to others. Its *Umwelt* "feels all-encompassing" because it is all that creature knows—and so is easily mistaken "for all there is to know."[4]

Even if we limit ourselves to our own species, no two people experience the world in exactly the same ways:

- Up to 8 percent of men and 0.5 percent of women have color vision deficiency (formerly called "color blindness"), most commonly making it harder to distinguish certain reds and greens.
- I happen to think cilantro is delicious, but an estimated 4 to 14 percent of people experience it as tasting like soap.
- And while body odor is certainly not my favorite smell, some people perceive it as smelling like vanilla.[5]

Reality is wild, y'all.

But these differences are not always minor. For many autistic people, sensory experience can be much more intense: a room most people would find unremarkable can feel genuinely overwhelming—its lights too bright, its background hum impossible to ignore. If human beings can experience the same world this differently, imagine how much greater those differences get across millions of species.

If we break the world down into the language of physics, "light is just electromagnetic radiation. Sound is just waves of pressure. Smells are just small molecules."[6] But the senses available to each species determine which aspects of that physical actuality can be perceived, and how those phenomena are experienced, interpreted, and acted upon. The physical world is one. The worlds that living creatures experience are many.

Yong's book explores how animals experience the world differently from humans through smell, taste, light, color, pain, heat, touch, water currents, vibrations, sound, echoes, electric fields, magnetic fields, and much more. Let's consider just a handful of examples.

Dog owners sometimes say, "We humans think we're taking our dogs for walks, but what our dogs really want is to be taken for smells." Anyone who has spent much time walking a dog knows this is true. You barely get started before the dog stops in its tracks, sticks its nose into some object, and proceeds to sniff it intently and at length. We have email; dogs have pee mail.

Here's why. Take a deep breath through your nose. As you inhale, notice that the same stream of air allows you both to breathe and to smell. Something similar happens when a dog inhales—but with one remarkable difference.

A dog's nose contains a separate passage devoted specifically to smell, diverting part of every breath into an olfactory chamber. As a result, when we humans exhale, we expel most of the odor molecules we have just detected. Dogs, however, retain those odor molecules in this specialized smelling passage, where each new sniff reinforces the last. For a dog, the world is mostly made of smells, not sights.[7]

Depending on the breed, a dog's sense of smell can be tens of thousands of times more sensitive than ours. Dogs can distinguish identical twins by scent, detect a single fingerprint left outdoors for a week, and be trained to locate everything from bombs and drugs to tumors and low blood sugar, and seizure-associated odors.[8]

Even though we humans share the same basic olfactory system as dogs, their olfactory world is far richer than ours, wider and deeper than anything we can readily imagine. The more I learn about the canine sense of smell, the more patient I become when walking a dog. I often don't have time to let them stop and investigate every intriguing scent, but I do try to give them more chances to sniff to their hearts' content.[9]

Although dogs tend to attract the most attention, many other animals (including rats, pigs, and elephants) are remarkable smellers in their own right. The beautiful film *Pig*, starring Nicolas Cage as a truffle hunter, offers a memorable glimpse into the scent-detecting abilities of pigs. The encouraging news is that, while we will never rival dogs, rats, pigs, or elephants—we're not in the same league—we humans can improve our sense of smell through intentional practice. Like so many aspects of perception, our senses are not entirely fixed. We can learn to pay closer attention to the worlds they reveal.[10]

We've begun with a sense in which humans are comparatively limited: smell. Now let's turn to sight, where we fare much better, though eagles and other birds of prey still see substantially more sharply than we do.[11]

Remarkable as our eyesight is, it also has a shadow side. Because vision dominates our experience, we easily assume that whatever catches our eye must also matter to other animals—and even to other people. Once again, our *Umwelt* (our "subjective universe," our "perceptual bubble") limits our ability to appreciate the many other ways of experiencing the same world.[12]

Even though we've shifted from smell to sight, let's stay with dogs for a moment. Many of us grew up hearing that dogs see only in black and white. It turns out that's a myth: "Dogs do see color. They just don't see the same range that people see." Instead, "they see mostly in shades of blue, yellow, and gray."[13] A red ball on green grass, for example, may be much harder for a dog to distinguish than a blue or yellow one. Evolution did not shape every species to perceive the world in the same way. Each sensory world reflects the particular opportunities of a creature's way of life.

This is where things start to get trippy: "Color is fundamentally subjective. There's nothing inherently 'green' about a blade of grass, or the 550-nanometer light that it reflects." The world contains wavelengths of light. Color is what nervous systems make of them. The colors a creature perceives depend on its sensory equipment. As one vision scientist quipped, "For a blue whale, the ocean is not blue." As best we can tell, whales perceive little or no color at all;

their visual world seems to be organized much more around light, contrast, and movement than around the hues familiar to us.[14]

If you've ever been to a horse race, you may have noticed that distracting human obstacles are often painted orange. To us, that wavelength of electromagnetic radiation stands out vividly. To horses, however, what we perceive as orange blends into the background. If we wanted to choose a color that horses would notice more readily, "we'd paint the marks fluorescent yellow, bright blue, or white."[15]

I hope even this brief tour of the senses has made the world a little less familiar to you, and a bit more interesting. The French novelist Marcel Proust captured this insight beautifully: "The only true voyage of discovery . . . would be not to visit strange lands, but to possess other eyes, to behold the universe through the eyes of another, of a hundred others, to behold the hundred universes that each of them beholds, that each of them is."[16]

The philosopher Thomas Nagel approached the same insight from a different angle in his essay "What Is It Like to Be a Bat?" He wasn't asking what it would be like for a human to have bat abilities (echolocation, flight, and so on). He was asking something harder: What is it like for a bat to be a bat? His conclusion was one of remarkable modesty: "If I try to imagine this, I am restricted to the resources of my own mind, and those resources are inadequate to the task."[17] Science can teach us an extraordinary amount about how bats perceive the world. The mystery of what that world feels like from the inside remains. Nagel is right about the limits of our awareness. Nevertheless, scientists continue pushing the boundaries of what we can learn about realities beyond our ordinary human senses. We may never fully inhabit another creature's perceptual world, but we can pay closer attention to it.

As Ed Yong writes:

> We may not ever know what it is to be an octopus, but at least we know that octopuses exist, and that their experiences differ from ours. Through patient observation, through the technology at our disposal, through the scientific method, and, above all else, through our curiosity and imagination, we can try to step into their worlds. We must choose to do so, and to have that choice is a gift. It is not a blessing we have earned, but it is one we may cherish.[18]

12

BEFORE ALLEGIANCE

What do I belong to before I choose?

At Standing Rock, the movement to stop the Dakota Access Pipeline became the largest Indigenous-led mobilization in decades. Over nearly a year, more than three hundred tribal nations came together, drawing global attention to Indigenous sovereignty, treaty rights, and the protection of land and water. Although the pipeline was ultimately completed, Standing Rock brought those struggles before a much wider public.[1]

Standing Rock also gave new force to a warning offered decades earlier by the legal scholar Felix Cohen: the treatment of Native peoples should be understood as the nation's "canary in the coal mine"—an early warning of dangers that may eventually threaten everyone else.

That insight has taken on new urgency in an age of

climate change and widening wealth inequality. As the Indigenous scholar and activist Dina Gilio-Whitaker writes, the possibility of a sixth mass extinction means that "We're all on the reservation now."[2] The ecological precarity and systemic injustices that Indigenous peoples have long endured increasingly confront us all.

Don't get me wrong—I'm all for creative solutions to mitigate climate change. But we're already living with its effects, and technological innovation alone cannot tell us how to endure, adapt, and live well as the world changes around us. Indigenous histories offer one sobering place to begin.

Historians estimate that the Indigenous population of what is now the United States declined by roughly 99 percent between European contact and the end of the nineteenth century: a demographic catastrophe that many scholars rightly describe as genocide.[3] Some Indigenous writers have therefore described the modern Indigenous experience as "post-apocalyptic": living in the aftermath of a world that has already ended.

Learning history from Indigenous perspectives is therefore both a warning and a source of resilience. It reminds us what societies lose when domination and extraction displace mutuality and care. It also points toward Indigenous traditions that have survived those forces and continued to understand human life through relationship, gratitude, and reciprocal obligation. That difference is not only about what we value. It is also about how we learn to see and know the world.

Robin Wall Kimmerer's *Braiding Sweetgrass: Indigenous Wisdom, Scientific Knowledge, and the Teachings of Plants* offers one of the clearest accounts of that way of seeing. The book seemed to find me before I finally found it. For several years, I kept hearing people mention it. Nearly everyone who recommended it added the same words: "I love that book." After finally reading it for myself, I understood why. It is a beautiful meditation on how scientific and Indigenous ways of knowing can enrich one another, and on what it means to live with reverence, reciprocity, and close attention to the more-than-human world. That matters far beyond how we understand plants. If we want to avoid repeating the patterns of domination and extraction at work in conflicts such as Standing Rock, we have to examine the habits of perception that make those patterns seem normal in the first place.

Kimmerer is a citizen of the Citizen Potawatomi Nation, a Distinguished Teaching Professor of Environmental Biology at the State University of New York, and the founding director of the Center for Native Peoples and the Environment. She describes herself as both a scientist and a poet.[4] For much of her life, however, she was told that she could not be both: that she would have to choose between those two ways of knowing.

From a scientific perspective, Kimmerer knows Canada goldenrod as *Solidago canadensis*, a member of the aster family. A botanical field guide can tell us its height, leaf structure, flowering pattern, and preferred habitat.

A plant can also be known poetically. Kimmerer describes a field of blooming goldenrod as an explosion of golden fountains standing beside the deep royal purple of New England asters: a dazzling display of complementary colors transforming an ordinary meadow into something almost regal.[5]

Yet poetry alone does not satisfy Kimmerer either. For her, wonder becomes a question. After describing the beauty of goldenrod and asters, she writes simply: "I wanted to know why."

Poetry and science need not be opposites. I recognize the impulse Kimmerer describes: beauty is often what makes me want to sit in a field and study flowers for hours, and learning more about what I see gives me even more to wonder about. But Kimmerer was taught to separate those ways of knowing. An academic advisor told her, "If you want to study beauty, you should go to art school."[6]

For many years, Kimmerer survived as an Indigenous woman in the predominantly white, male world of academic science by embracing only the scientific side of herself. But not long after earning her Ph.D., she was invited to join a gathering of Native elders to discuss traditional knowledge of plants.

There she listened as a Navajo elder spoke about plants for hours through a web of relationships: where each plant lived and when it bloomed. Which animals ate it and which birds lined their nests with it. What it offered and what it had to teach. Kimmerer realized she was encountering another careful and observant way of knowing plants. Reflecting on that encounter, she writes, "Her words were like smelling salts waking me to what I had known."[7]

The encounter helped Kimmerer see that the choice she had been taught to make was too narrow. She did not have to give up scientific rigor to take beauty, relationship, and her own place within the living world seriously. As Indigenous scholar Gregory Cajete has written, "In indigenous ways of knowing, we understand a thing only when we understand it with all four aspects of our being: mind, body, emotion, and spirit."[8] Kimmerer does not ask one way of knowing to replace another. She brings them into conversation, and that conversation changes what she notices and the questions she asks.

That integration reaches beyond how Kimmerer knows the world to how she understands her place within it and what she owes. Before we ever declare an allegiance, we are already shaped and sustained by relationships we did not choose. For Kimmerer, that tension surfaced early in the Pledge of Allegiance. As a child in public school, she recited it each morning, even as she struggled to reconcile its promise of "liberty and justice for all" with what she knew of her own people's history.[9]

Years later, the school called home because one of her daughters had stayed in her seat instead of standing for the pledge. Her explanation: "Mom, I'm not going to stand there and lie. And it's not exactly liberty if they force you to say it, is it?"[10]

The Pledge of Allegiance was introduced into American public schools in 1892 as part of the national observance of the four-hundredth anniversary of Christopher Columbus's first voyage to the Western Hemisphere.[11] At the time, Columbus was widely celebrated as a national hero. Today, many Americans recognize that his voyages also opened centuries of conquest, dispossession, and suffering for Indigenous peoples. That does not make ideals such as "liberty and justice for all" meaningless; they give us a standard for holding the nation to its own promises. As the bumper sticker puts it, "I love my country, but I think we should start seeing other people."

Kimmerer contrasts the daily recitation of the U.S. Pledge of Allegiance with the "Words That Come Before All Else." Also known as the Thanksgiving Address or the "Greetings to the Natural World," versions of it traditionally open and close important gatherings. It is an extended litany of gratitude for human community, the waters, plants, animals, winds, stars, and countless other gifts of the living Earth. Before any declaration of loyalty to a nation, it gives thanks for the relationships that make life itself possible.[12]

Kimmerer suggests that a culture that regularly begins in gratitude would live differently in the world: "While expressing gratitude seems innocent enough, it is a revolutionary idea. In a consumer society, contentment is a radical proposition." Gratitude shifts our attention from what we lack to what we've been given.[13]

Standing Rock and the Thanksgiving Address expose the same choice: whether we treat land, water, air, other living beings, and the Earth itself as objects to control or relationships to honor. That choice is built into systems as well as personal habits—in laws governing ownership and resource use, economic systems that reward extraction, and institutions that determine whose claims and whose futures matter.

Imagine starting each morning not by pledging allegiance to a country, but by naming the relationships that make your life possible: the water, the plants, the animals, the people. Do that for a year, and it might change more than your mornings. It might change you.

PART III

LIVING WISELY AS EARTHLINGS

Part I explored who we are. Part II introduced us to the more-than-human world with which we share the Earth. Part III asks what it means to live wisely as Earthlings.

Understanding is not enough. We may know that we are Earthlings. We may know that a forest is a community rather than merely a stand of trees, and that life is bound together in ways more intricate than we can easily picture. Wisdom begins when that knowledge changes how we live.

That change begins with attention, because what we notice shapes how we live. We become what we attend to. Even our deepest convictions struggle to take root in a hurried, distracted life.

The chapters that follow explore four ordinary disciplines: paying attention, exercising discernment, working at a human pace, and coming to our senses. None of them requires extraordinary talent. They require practice. Together they help us live more wisely as Earthlings.

13

WE BECOME WHAT WE ATTEND TO

What is my attention actually on?

Have you ever sat down first thing in the morning to work on something that truly mattered to you, only to find yourself repeatedly pulled away by emails, social media, text messages, phone calls, or people stopping by? Before you know it, it's noon. Your freshest hours of attention are gone, but the work that mattered most remains unfinished. After lunch, the interruptions continue, your energy begins to fade, and by mid-afternoon your deepest priorities have once again been crowded out by whatever demanded your attention most insistently.[1]

Most of us recognize this experience as a productivity problem. At its core, it is an attention problem. And because our lives are shaped by what we repeatedly attend to, it is, in no small measure, a life problem.

Our struggle to focus is intensified by the attention economy, which reaches us most pervasively through our smartphones. They keep the internet—and its endless stream of emails, texts, news, and notifications—close at hand. As a result, distractions can find us almost anywhere, interrupting not only our work but also our rest, our relationships, and our capacity to be fully present to the world around us.[2]

In *Attention Span: A Groundbreaking Way to Restore Balance, Happiness and Productivity*, Gloria Mark, professor of informatics at the University of California, Irvine, reports that people feel overworked and exhausted by information overload, that keeping up with email can feel Sisyphean, and that computers and smartphones make sustained focus increasingly difficult. Distraction also arises from within: from a thought, a memory, or the urge to look something up. As she puts it, "When you are immersed in the world's largest candy store, it is hard to resist sampling the wares."[3]

Mark's research reminds us that our attention is shaped not only by our own choices but also by the environments in which we live. Social media platforms, search engines, and many other online services are intentionally designed to capture and hold our attention for as long as possible. Endless scrolling, autoplay, personalized recommendations, and variable rewards all make it easier to keep clicking and scrolling. The result is more than lost productivity. We also become less able to give sustained attention to the people, places, and realities that matter most.

One practice has been transformative in my own life. For some of you, it alone might make the chapter worthwhile. I can testify that it works. There are weeks when I feel overwhelmed and my inbox overflows like everyone else's. Even so, I keep getting back to Inbox Zero. That doesn't mean every task is finished. It means every email has been processed and every commitment has been placed into a trusted system rather than left buried in the random order in which it arrived. Instead of rereading the same messages over and over, wondering what I still need to do, I know exactly what comes next.

The specific approach to Inbox Zero that has helped me most comes from productivity expert Carey Nieuwhof. At its heart is a remarkably simple workflow:

- Read each email only once.
- Then immediately decide what to do with it: respond if it takes only a few minutes, or archive it after recording the necessary next step on a trusted task list (I use a Google Doc) or scheduling time on your calendar to deal with it later.[4]

The goal is not an empty inbox for its own sake. The goal is to stop using your inbox as a to-do list. An inbox organizes messages by when they arrived, not by what matters most or what needs to happen next. Once every email has been processed into a reliable external system, you no longer have to waste time rereading the same messages to remember what still needs your attention. You free your attention for the work, the relationships, and the moments that deserve it.

Mark is right that Inbox Zero is never a permanent achievement. But for me, it is a recurring practice of creating mental space rather than chasing perfection.

If you're one of those people with hundreds, thousands, or even tens of thousands of emails sitting in your inbox, the first step is admitting you have a problem. You may eventually get to some of those old messages, but the likelihood drops sharply the longer they sit there. So declare email bankruptcy: archive the whole pile, start from a clean slate, and practice Inbox Zero from that day on. It feels unsettling for about a day. Then it feels like relief.

The same principle can preserve the benefits of a vacation after you return. You might set an auto-response like this: "I'm currently on vacation and won't be reviewing emails received while I'm away. To prevent an unmanageable backlog, please resend your message after [date]. Thanks so much!"

Helpful as practical systems like Inbox Zero can be, nothing has done more to strengthen my capacity for sustained attention than concentration meditation. Choose a single object of attention (such as your breath) and gently return your awareness to it whenever your mind wanders. Progress does not mean that thoughts, emotions, or physical sensations stop arising. It means becoming increasingly able to notice them without being carried away by them, gently returning your attention to your chosen object each time your mind wanders. Over time, you become less distractible, both on the meditation cushion and in everyday life.

If you'd like to try concentration meditation, pick a door.

- Behind Door #1, for an easy beginner's start: *Peak Mind: Find Your Focus, Own Your Attention, Invest 12 Minutes a Day* by Amishi Jha.

- Behind Door #2, for a more systematic approach: the early chapters of *The Mind Illuminated: A Complete Meditation Guide Integrating Buddhist Wisdom and Brain Science for Greater Mindfulness* by Culadasa (John Yates, Ph.D.).
- And behind Door #3, if you're curious about the intensely pleasurable states of deep concentration: *Right Concentration: A Practical Guide to the Jhanas* by Leigh Brasington.

Rather than simply reading about concentration meditation, let's practice it for a few moments right now. Like going to the gym to strengthen a muscle, we can strengthen our capacity for sustained attention through regular practice. There are many possible objects of attention. The most common is the breath, but people also focus on a candle flame, a statue, flowing water, or even a leaf or branch gently moving in the wind.

If you're willing, pause here and follow along as you read.

For the next few moments, practice being here—fully present with your whole heart, mind, body, and spirit.

Take a deep breath in...and out.

The Zen teacher Thích Nhất Hạnh might invite us to enter into this practice by becoming a "business-less person."[5] Imagine you have no business to attend to other than just sitting.

You're here. You showed up. Whatever happened before you arrived, and whatever awaits you after this meditation, there is nothing you need to do about it right now. For the next few minutes, experiment with having nothing on your to-do list except being here.

Nothing to do. Nowhere to go. Just be here, now, as each moment arises and passes away.

If your mind wanders, that's okay. Every time you notice you've become distracted, you've just experienced a moment of mindfulness. Rather than criticizing yourself, celebrate that noticing. Otherwise, you might have remained distracted much longer. Notice when your mind has wandered and—with kindness rather than criticism—return your attention to your breath.

Now let your attention widen to include your body. Notice where you feel the weight of your body against the chair or floor.

Feel your feet, your hands, your face.

See if there is anywhere you are holding unnecessary tension, and if so, let it soften without trying to force anything.

Then widen your attention once more. Notice sounds as they come and go.

Notice sensations, thoughts, or emotions arising and

passing. You do not need to hold on to any of them or push any of them away. For these few moments, simply let yourself be here, receiving whatever this moment brings.

Stay here as long as feels right to you.

When you are ready, continue reading, bringing some of this greater sense of presence with you. Notice whether the world feels even slightly different when you meet it with a little more attention.

Off the meditation cushion, however, presence is only the beginning. Life usually asks more of us than simply being present. When I think about all the competing demands on our attention, I'm reminded of the greased-watermelon contests we used to do at summer camp. You smear a watermelon with Vaseline, throw it into the middle of the swimming area, and everybody tries to wrestle it to a designated spot at once. Let me tell you: that watermelon is slippery. You think you've got a grip, and—whoosh—it pops right out of your hands. Trying to hold our attention steady amid the crashing waves of distraction can feel a lot like trying to hang onto a greased watermelon.[6]

Every time we shift our attention (from the project we're working on to an email, a text message, a phone call, or someone stopping by), our brains must construct a new mental framework for the task at hand. Imagine your mind as a whiteboard. Each time you switch tasks, your brain has to erase that whiteboard and sketch an entirely new mental map.

When we're interrupted repeatedly (or attempt to multitask), that whiteboard is constantly being erased and redrawn. Every switch carries a cognitive cost. Over time, those costs accumulate, leaving us mentally depleted. As Mark summarizes, constant task-switching contributes to "stress, fatigue, poor performance, and even burnout."[7] Attention is a finite resource. The more fragmented it becomes, the harder it is to offer our full presence to our work, to the people we love, or to the world around us.

Studies have found that the average amount of time office workers spend focused on a single screen before switching has steadily declined. In 2004, people stayed on one screen for an average of about two and a half minutes before shifting their attention elsewhere. By 2012, that average had fallen to seventy-five seconds. More recently, it has dropped to between forty-four and fifty seconds.[8]

Mark's research suggests that the average knowledge worker spends about ten minutes on a project before switching to email. From there, attention shifts to social media, then back to email, and then elsewhere again. Only about twenty-five minutes after that first interruption do they finally return to the original task. By then, their brain must rebuild the mental map it had already constructed. When this cycle repeats dozens of times each day, it becomes easy to see why so many people end the day exhausted, even when they feel they've accomplished surprisingly little.[9]

Mark's review of the research finds that frequent task-switching slows us down, increases errors, weakens performance on the main task, and raises stress.[10] Nor is this frenetic pace confined to the workplace. Our leisure has become increasingly fragmented as well. In 1950, the average shot in a television show or film lasted about twelve to thirteen seconds. By 2010, the average shot length had fallen to less than four seconds. Our entertainment, like our work, increasingly rewards rapid shifts of attention, conditioning us to expect constant novelty and making sustained attention feel increasingly unfamiliar.[11] When both our labor and our leisure train us to move continually from one stimulus to the next, sustained attention becomes increasingly difficult—and increasingly countercultural.

A related trend is "second-screening": texting while doing homework or scrolling through social media while half-watching a television show.[12] We have become so accustomed to dividing our attention that giving ourselves fully to a single activity can almost feel unusual.

One antidote is to watch better television. That doesn't mean everything we watch needs to demand our full attention. After a long day, I sometimes enjoy something light and undemanding too. But some works of art ask for—and richly reward—our complete attention.

Try watching the 2024 television series *Shōgun* without ever reaching for your phone. I defy anyone to follow that extraordinary series while simultaneously scrolling social media. You'll be lost almost immediately. Reading a long novel or sustained argument asks for the same capacity: the willingness to stay with something long enough for its complexity to unfold.

Or consider *Ripley*. A friend of mine who works in the film industry described it as "the most beautifully shot television series of all time, by some margin." You could pause at almost any moment, print the frame, and hang it on your wall as a work of art. Watching *Ripley* attentively is less like binge-watching television than wandering through a sumptuous black-and-white photography exhibit.

Reclaiming our attention doesn't always mean doing less. Sometimes it just means doing one thing at a time. The best experiences (a real conversation, a good meal, a piece of music you actually listen to) all ask the same thing of us: be here for it.

The internet has given us unprecedented access to information, creativity, and human connection. Its basic architecture of hyperlinks also makes associative movement unusually easy: one page leads to another, one source to another, one idea to the next.[13] That can be tremendously useful, helping us follow curiosity, clarify confusion, or go deeper into a subject. But it also makes attention easy to redirect.

Mark argues that online environments can be especially distracting because this associative structure resembles the way thoughts connect in the mind. The problem is not hyperlinks themselves; a link can function much like a footnote or citation, quietly offering a path for anyone who wants to follow it. The attentional pressure becomes much stronger when links are combined with notifications, recommendations, autoplay, advertising, and other features designed to capture and hold our attention.

The answer is to inhabit the internet more wisely. For many of us, that means becoming more intentional about when, why, and how we engage with it, so that we direct our attention rather than letting platforms direct it for us.

Personally, I've found it fairly easy to say, "Get off my lawn, TikTok!" But I also know plenty of people who enjoy TikTok or learn a great deal from it. My own digital temptations lie elsewhere: I still catch myself checking Facebook more often than I intend.

Whatever you use, it's worth remembering that on major commercial social media platforms, keeping us engaged is central to the design. Algorithms and other features are optimized to keep us watching, scrolling, or clicking because our attention is valuable to the companies that capture it. Choosing where to direct our attention therefore becomes both a personal discipline and a form of resistance to systems built to profit from it.[14]

The difficulty is only going to grow as virtual reality and other immersive technologies become increasingly sophisticated. Steven Spielberg's *Ready Player One* imagines a future in which many people prefer spending their lives in virtual worlds to engaging with the world around them. Whether or not the future looks exactly like that, the film asks us: To what kind of reality do we want to give our attention? For where we repeatedly direct our attention is, over time, the world we inhabit—and the kind of people we become.[15]

Three practices can help us reclaim our attention.

First, protect your best attention. Although there are dozens of time-management systems, many of the most respected approaches converge on one central insight: devote one or more of your first and freshest hours of the day to your single most important task before opening email, checking social media, or responding to less important demands.[16] In other words, don't let the urgent crowd out the important. Our scarcest resource is not time but attention, and where we give our best attention reveals—and gradually shapes—what matters most to us.

Second, invest your best attention in your closest rela-

tionships. Psychologist Robin Dunbar argues that human beings can sustain only a limited number of meaningful relationships:

- about 150 people in our broader social circle
- about 12–15 people with whom we stay meaningfully connected through at least monthly contact
- only about 3–5 truly intimate relationships that we nurture week by week

Social media may tell us we have thousands of "friends," but that isn't true in any meaningful sense. Our attention is finite. There simply isn't enough time—or emotional energy—to build deep relationships with everyone.[17]

Building on this insight, Carey Nieuwhof offers a helpful leadership principle: "The depth of the relationship should determine the depth and speed of your response."[18] Imagine how much calmer, and how much less fragmented, your life might become if you responded less urgently to people on the outer edges of your social circle while giving more intentional attention to those closest to you.

Third, practice giving your full attention to whatever is before you. Whether you're reading a great book, sharing a meal with a friend, walking through the woods, or sitting quietly with your own breath, resist the temptation to divide your attention. Life's richest experiences rarely happen while we're half somewhere else. They ask for our full attention.

Jenny Odell, an artist and writer who teaches at Stanford University, offers an important next step in *How to Do Noth-*

ing: Resisting the Attention Economy. Despite its provocative title, the book is not really about doing nothing. As she explains, "The first half of 'doing nothing' is about disengaging from the attention economy; the other half is about reengaging with something else."[19] Reclaiming our attention, then, is only the beginning. The deeper question is what deserves our attention in the first place. Living wisely as Earthlings requires more than knowing that we belong to a living world. We have to give enough attention to that world—and to one another—for that belonging to shape how we live.

What, then, is worthy of your attention? What practices, relationships, or places help you become more fully alive? What habits are no longer serving you? What might become possible—for you, for others, or for the wider world—if you reclaimed more of your attention from the attention economy?

Our attention is finite, and once a moment is gone, we don't get it back. Attention is how we spend our lives. We become what we attend to.

14

NEVER SAY NEVER, NEVER SAY ALWAYS

What am I calling normal that I just prefer?

When you hear the word natural, perhaps you picture a hidden waterfall deep in a forest, a peaceful grove of trees, or a sweeping mountain vista. But natural does not necessarily mean peaceful, healthy, or good. Hurricanes are natural. Earthquakes are natural. So are viruses, parasites, and epidemics.

Learning to live more consciously as Earthlings can strengthen our respect for the living world. It can also tempt us to romanticize nature. Belonging requires seeing the world as it is, not turning "natural" into another word for "good."

The COVID-19 pandemic made that distinction vivid. The natural course was for the virus to spread; masks, distancing, and vaccines were deliberate interventions in that process. People still argue about which of those interventions were wise—but that argument is exactly the point. "Natural" by itself cannot tell us whether a response is good. Sometimes the wisest response to nature is to intervene.

Nevertheless, natural and unnatural remain among the most rhetorically powerful words in our language. They often function less as careful descriptions than as tools of persuasion. Before going further, it is worth taking a closer look at these deceptively simple words and discovering how to recognize when they are being used to manipulate rather than clarify.

All too often, we are presented with a false dichotomy. Natural comes to mean everything healthy, authentic, pure, and good. Unnatural, in turn, becomes synonymous with everything unhealthy, artificial, corrupt, or wrong. The world, however, is rarely so simple. As is often the case, the truth is messier and more complex than either label suggests.

This insight is hardly new. In *Natural: How Faith in Nature's Goodness Leads to Harmful Fads, Unjust Laws, and Flawed Science*, religion professor Alan Levinovitz explores it at length. More than a century earlier, the philosopher John Stuart Mill similarly observed that "Nature, natural, and the group of words derived from them...are one of the most copious sources of false taste, false philosophy, false morality, and even bad laws." Indeed, lists of common logical fallacies almost always include the "appeal to nature."[1] Labeling something natural tells us surprisingly little. The better question is whether something helps us live wisely and well.

History offers countless examples of appeals to nature

being used to justify systems of oppression. For generations, women were told they were naturally inferior to men. In a classic case of victim blaming, misogynists argued, "It's not that we want women to be second-class citizens; it's simply the way nature intended things to be." So you bar women from universities and professions, deny them political power, and then point to men's greater achievements within that rigged system and say, See? That's just how nature wanted it.

We could trace a similar pattern in the histories of white supremacy and eugenics, or in the once-common claim that interracial marriage was unnatural, despite the fact that race is a social construct rather than a biologically discrete reality: at the DNA level, any two humans are about 99.9% identical.[2]

Scratch beneath the surface of many appeals to what is supposedly natural or normal, and you will often find one group attempting to universalize its own prejudices. Sexism, racism, homophobia, classism, and ableism are recast not as human choices, but as the inevitable order of nature.[3]

For many socially progressive readers, everything I've said so far is 101-level stuff: the relatively easy work of unpacking the natural/unnatural binary. So let me turn the dial up a notch. We're about to wade into some turbulent waters, and I'll say up front: my aim isn't to come down too hard on one side of the debates I'm about to name. It's to notice, in each case, how the words natural and unnatural are doing the arguing for us, and how recognizing that hidden argument opens up a wider range of possibilities.

Take the case of genetically modified food, sometimes disparaged as "Frankenfood." It is far more useful to ask what the best available scientific evidence says about its risks and benefits. Might a particular modification reduce pesticide use, help feed more people, lessen malnutrition, or make crops more resilient to drought? Or might it introduce new environmental, economic, or health risks? Those are issues worth debating.

Or take the controversy over "natural childbirth." Take a deep breath—we'll get through this. Few topics run hotter. Parents, physicians, midwives, doulas, and childbirth educators disagree passionately, and understandably so. But the questions that actually help are about the health of parent and child, not whether an intervention gets labeled natural or unnatural.

Imagine two different situations. In a well-resourced health-care system, the standard birthing process may at times become overly medicalized, making it worthwhile to question routine interventions. I've spoken with many people who had to advocate strongly to create the birthing environment that was right for them. But in a part of the world with high maternal and infant mortality rates, advocating for greater access to those very same interventions, even if they are labeled unnatural, may be exactly what is needed.[4] The point is not to decide in the abstract whether an intervention is natural or unnatural, but to ask what this particular situation actually requires.

That also means resisting the temptation to romanticize what is natural or to assume that there is a single "normal" appropriate for all people, times, and places.[5] As Levinovitz writes:

> "Natural" is not synonymous with beautiful, healthy, true, or good. Once freed from this...understanding of nature, it becomes possible to discuss the potential benefits for natural childbirth—and there are many—without evangelizing obedience to nature. It also becomes possible to state clearly the benefits of unnatural childbirth—and there are many—without being labeled a heretic.... Then a new space will open up where nature can be seen for what it really is: one among many potential sources of knowledge and value, which differs in importance from culture to culture, or from person to person.[6]

Many of these issues are better understood as "polarities to be managed" than as "problems to be solved."[7] People may land at different points along the spectrum depending on their circumstances and season of life. Wisdom lies not in choosing one pole once and for all, but in discerning what is life-giving in a particular context.

From 2000 through 2024, the United States usually recorded fewer than one hundred measles cases a year; even the previous high, in 2019, was 1,274. In 2025, the country recorded 2,289 cases. By August 13, 2026, it had recorded 2,566 more—surpassing the previous year's total before the year was over. Measles is not a minor childhood illness: it typically brings high fever, cough, red, watery eyes, and a rash that spreads across the body, and it can lead to complications such as pneumonia and brain inflammation. Three unvaccinated people, including two school-aged children, died of measles in 2025.[8] Sometimes the natural process is the disease that can kill us, while the intervention labeled "unnatural" is what protects us from it.

Prescribed forest burns offer another example. For decades, putting out every wildfire looked like the obvious way to protect people and forests. Then we learned that many ecosystems evolved *with* fire. In some places, a carefully managed burn—which sounds about as "unnatural" as it gets—is exactly what restores an ecosystem's health. Again the useful question isn't natural-or-unnatural. It's what actually helps a particular place thrive.

Finally, a personal example. Every morning, the first thing I do is take Synthroid, a synthetic form of thyroid hormone. I depend on it because half of my thyroid was surgically removed, and the remaining half produces very little hormone on its own. Both surgery and synthetic hormones are deliberate interventions in what my body would otherwise do, and I'm very grateful for both—because if the only option were to let nature take its course, I wouldn't be here.

Once we recognize that natural is not always good and unnatural is not always bad, we become less susceptible to manipulative rhetoric and freer to weigh the full range of possibilities against the best evidence we have. We can then choose what seems wisest for ourselves and, when it matters, for our communities. And we can extend that same freedom to others, trusting that thoughtful, well-informed people will sometimes reach different conclusions than we do.

What matters is not simply whether something is natural, but whether it helps living beings flourish: whether it reduces suffering, advances justice, and supports life. Those questions rarely yield simple answers. But they move us past slogans and into the harder work of discernment that a complex, changing world requires.

Levinovitz captures that posture in a memorable line: "Never say never; never say always." There are certainly times I'd prefer more certainty. But more often than not, rigid certainty lands us in ideology and dogmatism. Uncertainty isn't always comfortable, but being willing to revise our views when the evidence requires it is essential to liberal religion.

Living wisely as Earthlings requires that same willingness: refusing to let the labels natural and unnatural decide for us. We belong to the living world, but belonging does not make nature a moral rulebook. We still have to look closely, weigh the evidence, and decide how best to live as part of the world we actually inhabit.

15

TWO WEEKS ON A PICNIC TABLE

What is enough?

Modern life often feels as though it is accelerating beyond our capacity to keep up. We answer emails while eating lunch, listen to podcasts while exercising, scroll through our phones while waiting in line, and measure our worth by how much we manage to accomplish. But despite all our technologies for saving time, many of us feel as though we have less of it than ever.

Cal Newport's *Slow Productivity: The Lost Art of Accomplishment Without Burnout* helped me rethink not just how I work, but the pace at which I want to live.

For one of my favorite case studies from his book, travel back with me to the summer of 1966. John McPhee was already a successful journalist. After seven years as an associate editor at *Time*, he had become a staff writer at *The New Yorker*, where he had already published five long-form articles. Despite those accomplishments, he found himself struggling with his most ambitious assignment: weaving together the geological, ecological, and political history of the New Jersey Pine Barrens.[1]

McPhee had already invested eight months researching the Pine Barrens but still could not figure out how to begin writing. For nearly two weeks, he spent his so-called working hours lying on a picnic table, "staring up into branches and leaves, fighting fear and panic." How many of us have bosses who'd be fine with us spending two weeks daydreaming while horizontal?

What looked like idleness was an essential part of the creative process. By giving himself permission to let his mind wander for an extended period, McPhee discovered the organizing structure he needed to weave together his vast body of research. The result became one of the most admired works of his career.[2]

How many magazine articles are published every month? A lot. How many are still read and admired decades later? Almost none. From the standpoint of modern productivity culture, spending two weeks doing nothing is a scandalous waste of time. And plenty of managers have been taught that good work "requires increasing busyness—faster responses to emails and chats, more meetings, more tasks, more hours."[3] But more hours in the office do not necessarily produce better results.

If you've seen the classic 1999 workplace satire *Office*

Space, you may remember the scene in which the protagonist cheerfully explains that he arrives late, slips in through a side door so his boss won't notice, spends much of the day staring at his desk while appearing to work, and estimates that he does only about fifteen minutes of actual work in a typical week.

In *Slow Productivity*, Newport criticizes this all-too-common emphasis on appearing busy, calling it "pseudo-productivity...the use of visible activity as the primary means of approximating actual productive effort."[4] Instead of rewarding people merely for looking busy, he asks how organizations can foster work that produces meaningful results. The research suggests that some of our best work requires something modern workplaces often discourage: giving people the time and space to think deeply, solve difficult problems, and create work of lasting value.

In 1995, the television network CBS was struggling in the ratings. One Friday afternoon, the head of entertainment noticed that the offices were mostly empty. He stormed back to his office and fired off an angry memo: "Unless anybody hasn't noticed, we're in third place [in the ratings]. My guess is that at ABC and NBC they're still working at 3:30 p.m. on Friday. This will no longer be tolerated."[5]

The memo captures exactly the confusion Newport is describing. It tells us nothing about how many hours CBS employees actually worked compared with their competitors. What it reveals is the assumption that an empty office meant people weren't working hard enough—and that more visible activity would produce better results.

A few years later, CBS offered a different picture of

creative work. Anthony Zuiker spent more than three years developing *CSI*, with long stretches when little visible progress was being made, punctuated by periods of intense creative work. The series went on to run for fifteen seasons and became one of the most successful television dramas of its era.[6] Once the show became a phenomenon, the years it had taken to develop hardly looked like wasted time.

Slow work does not guarantee success. But meaningful creative work can take time, and it does not always look productive from the outside. Busyness is not the same as productivity, and a frenetic pace does not necessarily produce work of lasting value.[7]

Inspired by movements such as Slow Food, which favors leisurely communal meals rooted in local and seasonal traditions; Slow Medicine, which gives caregivers time to attend to the whole person rather than focusing narrowly on disease; Slow Schooling, which resists high-stakes testing and competitive tracking in favor of richer learning; and Slow Media, which promotes slower, higher-quality alternatives to digital clickbait,[8] Newport coined the term slow productivity, defining it as:

> A philosophy for organizing knowledge work efforts in a sustainable and meaningful manner, based on three principles:
>
> Do fewer things.
> Work at a natural pace.
> Obsess over quality.[9]

Those three principles provide a helpful framework for our work and for living a more fully human life.

Before we explore these principles, it's important to acknowledge that many people have far too little control over their working hours or assigned tasks, and some workplaces actively reward the very culture of busyness Newport is criticizing. Even so, these ideas can be applied to areas of life where we often have greater freedom (our family commitments, hobbies, personal projects, and other priorities) to make life more humane and sustainable.[10]

Newport's first principle of slow productivity is to "do fewer things." Reduce your obligations until you can easily imagine finishing them with time to spare, then pour that freed-up energy into the small number of projects that matter most.[11] Or, as an internet meme puts it even better:

> Pick your battles.
> Pick fewer battles than that.
> Put some battles back.
> That's still too many battles.

Along these lines, I sometimes ask myself how I might become 80 percent busy instead of 120 percent busy. As Steve Jobs once put it, "Deciding what not to do is as important as deciding what to do."[12] How might your life benefit from doing fewer things so that you can bring your best self to the things that matter most?

Here's another example. Jane Austen is renowned for a remarkable five-year period publishing *Sense and Sensibility* (1811), *Pride and Prejudice* (1813), *Mansfield Park* (1814), and *Emma* (1816).[13] After her tragically early death at the age of forty-one, a romantic myth took hold: that she had written these masterpieces by "writing secretly, scribbling prose in bursts between the many distracting obligations of her social standing."[14]

The actual situation was quite different. In 1809, two years before publishing her first novel, Austen's family moved away from the hubbub of society to a quiet cottage where they enjoyed much greater privacy. There, Austen's only regular household responsibility was preparing breakfast for her family. Otherwise, she was free to write. In other words, Austen's greatest literary achievements emerged after she gained the time and freedom to devote herself more fully to her craft.[15]

Newport notes that Tolkien's great imaginative work followed a similar pattern. Much of what eventually became *The Lord of the Rings* took shape only after illness unexpectedly granted him extended periods of uninterrupted time.[16]

My own experience has reinforced the same lesson. Much of the writing of this book became possible only because my congregation granted me a three-month sabbatical. Stepping away from the constant rhythm of sermons, meetings, emails, and ministerial responsibilities created the long stretches of time needed for sustained reflection, writing, and editing. Like Austen and Tolkien in very different circumstances, I was reminded that some kinds of work take shape only when given enough time.

Newport's second principle is to "work at a natural pace." Instead of rushing our most important work, he encourages us to give it a sustainable timeline, recognizing that creativity often requires alternating seasons of intense effort, reflection, and renewal.

As background, Newport studied the lives of some of history's most influential scientists and noticed a striking pattern. They were extraordinarily productive, but by modern standards they often worked at an uneven—and sometimes surprisingly leisurely—rhythm.[17] Great work, it turns out, often takes longer than our productivity-obsessed culture expects.

Copernicus, for instance, developed his heliocentric model over many years. By 1514, he was circulating a brief outline among colleagues. He continued developing and refining the work for nearly three decades before publishing *On the Revolutions of the Heavenly Spheres* in 1543. Five centuries later, we are still talking about it.

Isaac Newton followed a similar pattern. He began thinking seriously about gravity as a young man, spent roughly fifteen years developing his ideas about the inverse square law, and then waited another fifteen years before publishing his theories.[18]

Here's a more contemporary example. A persistent myth claims that Jack Kerouac wrote *On the Road* in just three weeks. In actuality, that famous three-week typing spree came only after roughly three years of development, followed by another six years of revision before publication.[19] The legend celebrates the sprint, not the nine years of patient craftsmanship that made it possible.

The modern pressure to grind out more and more work, as if we were machines instead of human beings, is often counterproductive to creativity at the highest levels.[20] This is the heart of Tricia Hersey's fierce and subversive book *Rest Is Resistance: A Manifesto*. Hersey, who calls herself the "Nap Bishop," doesn't mince words: "Grind culture wants us to keep going no matter what. I sit my ass down and daydream. The answer is NO."

Newport suggests three practices for creating more time and space in your life. These particular examples may not work for everyone, but they can inspire variations that fit your own circumstances.

Practice 1: Try a No-Meeting Monday. (Or choose a different day if that better fits your schedule.) You may have heard of Meatless Mondays. Why not experiment with No-Meeting Mondays? One day a week amounts to only 20 percent of your working time, so you may find you can protect that time without giving the impression that you're unreasonably unavailable.[21] The payoff is a regular block of uninterrupted time for deep work.

Practice 2: See a matinee once a month.[22] I appreciate that Newport sets the bar at monthly rather than weekly. If movies aren't your thing, choose something else. Could you carve out even a few hours each month to do something genuinely restorative?

Practice 3: Schedule rest projects.[23] We've become accustomed to adding more and more items to our to-do lists. What if we became just as intentional about scheduling periods of rest, especially after particularly demanding seasons?[24]

Newport's third principle is to "obsess over quality." He argues that producing exceptional work sometimes requires passing up short-term opportunities, trusting that excellence itself will create greater freedom and more meaningful opportunities over time.[25]

Here's what that can look like. In 1995, Jewel was living out of her car and trying to make it as a recording artist. When a record label offered her a million-dollar signing bonus, she shocked them by turning it down.[26] She understood that so-called "bonuses" were really advances against future record sales, and she realized that accepting less money would buy her greater patience from the record label if success didn't come immediately. By accepting less money up front, she would leave the label with less to recoup—and therefore less pressure to become an immediate commercial success. That decision gave her the time and freedom she needed to hone her craft.

She reinforced that freedom by keeping her expenses low. Instead of hiring a tour manager and traveling on a tour bus, she drove a rental car and performed without a band. She even accepted daytime performances at high schools in exchange for free transportation.

Freed from the pressure to produce immediate commercial success, she gradually found her artistic voice. Her music became looser and more soulful. When she released a new video for "You Were Meant for Me" that reflected this more authentic style, her career took off. Sales grew from only a few thousand copies during its first year to roughly a million copies each month, and the album eventually sold more than twelve million copies.[27]

What might it look like to experiment with a slower, more intentional pace in this season of your life? Are there ways you can free yourself or others around you from pseudo-productivity?[28] How might your life benefit from doing fewer things, working at a more human pace, and obsessing over quality?

Part of living wisely as Earthlings is accepting that we are finite creatures, with only so much time and attention to give. Slowing down allows you to pour your best, most focused effort into the work that actually matters.[29]

Newport ends with a plea rather than a prescription: step back from the frenzy. Small, steady drops of real effort, gathered day after day across a year, add up to a bucket that's pretty full. As Newport puts it, "The time has come to try something slower."[30]

16

COMING TO OUR SENSES

What is my body telling me?

Mindfulness has undergone a remarkable cultural change over the past few decades. Once largely associated with Buddhist practice, it has become part of mainstream conversations about mental health, education, medicine, leadership, and well-being.

Jon Kabat-Zinn, a molecular biologist who helped bring mindfulness into mainstream Western medicine, defines it as "paying attention in a particular way: on purpose, in the present moment, and nonjudgmentally."[1] Mindfulness was the first form of meditation I practiced consistently, and I have experienced firsthand many of its benefits. So have millions of others.

Mindfulness is only one way of becoming more fully present. Two complementary dimensions are heartfulness and bodyfulness. No single approach is inherently superior. Temperament, experience, and season of life all shape which forms of contemplative practice prove most helpful. Just as physical cross-training develops different strengths, "contemplative cross-training" can broaden our human capacities.[2]

Some of the best-known heartfulness practices include loving-kindness, compassion, equanimity, and empathetic joy. They train qualities of heart we may want to strengthen. In loving-kindness practice, for instance, we silently offer wishes for well-being to ourselves and others; compassion practice asks us to turn toward suffering with care rather than judgment or avoidance. I have found the practices in Kristin Neff and Christopher Germer's *The Mindful Self-Compassion Workbook* particularly valuable.[3] They have taught me that the way we relate to ourselves in moments of suffering shapes every other aspect of our lives, including our relationships with other people and with the wider world.

Heartfulness remains less familiar than mindfulness, and bodyfulness less familiar still. Bodyfulness encourages us to listen to the wisdom of the body: to recognize that our bodies often perceive, remember, and communicate things long before our conscious minds catch up.[4]

I learned much of what I understand about embodied meditation and bodyfulness from the teacher Reggie Ray. His book *Somatic Descent* offers one of the clearest introductions to embodied awareness I have encountered.

Notice how often bodily awareness already shows up in everyday language:

- "What she said felt really heavy": sensing weight in the body.
- "So-and-so's behavior makes me sick": awareness of nausea.
- "I have a gut feeling about this": the body registering something before conscious thought.
- "I had a queasy sense that he was lying": awareness of unease.[5]

Expressions like these show how bodyfulness begins in ordinary experience: by paying attention to what our bodies may already be trying to tell us.

We can also deepen that awareness through one of Reggie Ray's practices.

A good place to begin is by assuming a comfortable, relaxed posture—or even lying down, a longstanding but often underappreciated meditation posture.

Then gently open your awareness to your body and silently ask, "What's actually going on with me right now?" Pause, and notice what arises.

Is there tightness in your gut, your chest, or your neck and shoulders? If so, what might it be trying to tell you?

You might then ask, "What needs my attention right now?[6]

Don't force an answer. Remain curious and receptive, allowing it to arise in its own time.

To practice bodyfulness more fully, we also need to reckon with the cultural forces that teach us to distrust, criticize, and disconnect from our bodies. One of the most powerful is what the artist, activist, and educator Sonya Renee Taylor calls the "Body Shame Profit Complex."[7] In purely economic terms, recent market analyses estimate the global beauty and personal care sector in the hundreds of billions of dollars annually, with forecasts passing the $1 trillion mark in the coming decade.[8] There's an enormous amount of money to be made in convincing us our bodies aren't good enough.

Taylor develops this critique in *The Body Is Not an Apology: The Power of Radical Self-Love.* Healing our relationship with our bodies, she argues, isn't just a matter of personal self-esteem. It's an act of both personal and social renewal.

Taylor's work grew out of an unexpected conversation. At the time, she had no intention of starting a movement that would eventually reach hundreds of thousands of people. Instead, she was listening to a friend who feared she might be unexpectedly pregnant. Her friend confessed, "My disability makes sex hard already.... I just didn't feel like it was okay to make a big deal about using condoms." In that moment, Taylor found herself speaking six words that would transform not only her friend's life, but also her own and, eventually, the lives of countless others: "Your body is not an apology."[9]

Today, Taylor regularly gets photos from people who've had "The Body Is Not an Apology" tattooed on their skin as a permanent reminder. That's one way to make a commitment to bodyfulness. Those six words turn bodyfulness into a commitment to treat ourselves, and our bodies, with more compassion and respect.

The project reaches beyond body positivity to examine the judgments we make about ourselves and others based on their bodies. Taylor calls us to loosen the grip of body shame and grow toward love and celebration of all the ways bodies differ, very much including our own.[10]

Three core principles are especially important:

1. Make peace with not understanding.
2. Make peace with difference.
3. Make peace with your body.[11]

Together, these three principles call us toward greater compassion and acceptance for ourselves, one another, and bodies in all their diversity. Let's take each in turn.

The first is to make peace with not understanding. Many of us gravitate toward living primarily in our heads. Cognitive intelligence has tremendous value. But the heart and body can register what conceptual thought misses.

There is so much we don't know about ourselves, much less about other people. Recognizing those limits can make us more compassionate, more curious, and less quick to judge. Instead of jumping to conclusions about another person based on how their body looks, what if we tried "accepting humans and their bodies without understanding why they love, think, move, or look the way they do?"[12] Can you feel some peace descending with that?

In Buddhism, this posture is sometimes called "beginner's mind": setting aside preconceptions in order to encounter the world with openness, curiosity, and freshness. Notice how these different ways of being show up in your body. Shame and judgment often feel tight, constricted, and disconnected. Love, curiosity, and acceptance tend to feel more open, connected, peaceful, and at ease.

Pause for a moment and recall a time when you felt deeply ashamed or harshly judgmental.

Notice what happens in your body as you remember it.

Where do you feel tightness, heaviness, heat, or contraction?

Then imagine extending love toward yourself or someone else.

Notice what changes in your body, even slightly. Is there any softening, warmth, spaciousness, or ease?

Stay with that difference for a moment.

Noticing how differently judgment and openness feel in the body is itself an act of bodyfulness. It can help us remain curious about what we do not yet understand rather than rushing toward shame or certainty.

Second, make peace with difference. Here Taylor draws on Audre Lorde's call to celebrate human difference rather than merely tolerate it.[13] The Body Shame Profit Complex tries to convince us that there is a single standard of beauty and that the more we deviate from that supposed norm, the less worthy we are. What a tremendous source of freedom it can be to lay down that oppressive standard and embrace the wild, beautiful diversity of human bodies—including our own.[14]

But making peace with difference asks more of us than simply admitting that bodies differ. It means noticing how we sort those differences into hierarchies: beautiful and unattractive, normal and abnormal, desirable and undesirable, healthy-looking and unhealthy-looking. Many of those judgments arrive before we have consciously chosen them.

Bodyfulness gives us a chance to notice that reflex: to feel judgment arise without immediately believing it. ("You don't have to believe everything you think!")[15] From there, we can practice meeting another body, and our own, with curiosity instead of a verdict. As we do, we become more at home in our own bodies and more able to celebrate the bodies of others. Making peace with difference, it turns out, is another way of making peace with ourselves.

Third, make peace with your body.[16] In a brief poem titled "Three," Nayyirah Waheed imagines a person trying to repair a long-strained relationship with their own body. It captures how many of us spend years treating our bodies as adversaries—and how much may change when that hostility gives way to care.

As you sit with that possibility, notice what happens in your own body. Is there even the slightest sense of softening, openness, or peace?

Right now, place one hand over your heart and the other

over your abdomen. Offer yourself a gentle, compassionate touch.

Take a slow, deep breath in…and out.

Listen for what your body may be telling you.

Rest there for a moment.

Through our bodies, we belong to the living world.

PART IV

BECOMING GOOD ANCESTORS

The first three parts of this book trace an expansion of perspective:

Part I asked us to see ourselves more truthfully—as Earthlings: participants in a vast, evolving universe and members of a living world, rather than as isolated individuals or masters of creation.

Part II looked beyond the human world. We considered the hidden lives of trees, fungi, plants, and animals, and found the living Earth stranger and more intelligent than most of us were taught.

Part III turned from understanding to practice: how attention, discernment, embodiment, and healthier rhythms can help us live in that larger world more wisely.

Part IV widens the horizon again: from how we live now to what our lives make possible for those who come after us. How do we face climate change, biodiversity loss, democratic backsliding, and the other upheavals of this century without giving in to cynicism or despair? What do we owe one another? What do we owe future generations? And what kind of spirituality can sustain us as we take up these tasks together?

The chapters that follow take up those questions through four themes: living through ecological crisis, hope without illusion, becoming good ancestors, and the future of liberal religion. They point toward something more demanding than optimism. Optimism assumes it will all work out on its own. What Buddhist scholar and activist Joanna Macy calls "active hope" begins with the conviction that the outcome is still open, and that what we do matters.[1] That work begins in the communities we build, the institutions we strengthen, and the habits we form today.

To belong to a living world is also to owe something to those who will inherit it. Becoming good ancestors is one way of taking that seriously. New technology and better policy will be essential. But we also hold this blue planet in trust for generations we will never meet. What those generations inherit will be shaped in part by what we do now.

17

COMMUNITIES WORTH CHOOSING

What am I really choosing when I choose where to belong?

In college, I double-majored in religion and philosophy. Decades later, I remain captivated—not only by religious texts and ideas, but by the living, evolving realities of religion and spirituality around the world. Every year brings fresh reminders that religion is anything but static. New movements emerge, ancient traditions adapt, old conflicts persist, and unexpected forms of spiritual life take shape. Religion is not a relic of the past. It remains one of the most dynamic forces shaping our world today.

Here are just a few snapshots from one recent year in the world's religious life:

- January in Istanbul: Greek Orthodox Christians dive into the Golden Horn to retrieve a wooden crucifix during Epiphany, carrying forward a tradition that has endured through empire, modernity, and change.
- February in East Asia: Families visit temples during the Lunar New Year season, lighting incense and offering prayers for blessing, renewal, and good fortune.
- March in northern Gaza: Palestinian Muslims come together for Eid al-Fitr prayers marking the end of Ramadan, sustaining religious fellowship amid hardship and hope.
- April in Vatican City: Hundreds of thousands of mourners gather in St. Peter's Square to honor Pope Francis, illustrating how ritual, memory, and grief can unite people across cultures, nations, and generations.
- May in Nepal: Buddhist pilgrims circle the great stupa at Boudhanath during Vesak, celebrating the birth, enlightenment, and death of the Buddha through prayer, lamps, and quiet acts of devotion.
- June in Canada: Sikhs fill the streets in colorful Nagar Kirtan processions, sharing free meals with neighbors and strangers alike, a vivid expression of their commitment to equality, hospitality, and service.
- July in Iraq: Muslims gather ahead of Ashura, the holiest day of the Shi'ite calendar, lighting torches during the sacred month of Muharram.

- August in India: Jain pilgrims observe Paryushana, a season devoted to fasting, self-examination, forgiveness, and reconciliation.
- September in New York City: Jewish congregations welcome Rosh Hashanah with prayer, reflection, and the sounding of the shofar.
- October in Bangladesh: A child reaches toward an oil lamp during Diwali, where light, wonder, and tradition are passed from one generation to the next.
- November in New York City: A priest prays as ICE officers prepare for immigration detentions at federal court, bearing witness to how religious ritual enters spaces of power, fear, and contested justice.
- December in Australia: A menorah is projected onto the sails of the Sydney Opera House after an antisemitic terrorist attack on a Hanukkah gathering—public space turned into collective mourning, a stand against hatred. May the memory of all those who were killed be a blessing.[1]

One of these snapshots deserves a closer look, because it complicates any simple story about religion and violence. The attackers were inspired by Islamic State ideology. The man who tackled and disarmed one of the gunmen was also a Muslim. The problem is not religion itself, nor any one tradition, but what happens when a strand of any religion hardens into the kind of fear, domination, and violence that it then calls holy.[2]

Individual stories matter because statistics cannot show us what religion feels like in a human life: how it shapes grief, courage, belonging, prejudice, resistance, and hope. But stories alone cannot tell us how the religious landscape is changing. For that, we need to turn to the data.

Because this book is written from within the liberal religious tradition, I want to focus on one part of that larger picture: how liberal religion is faring in the United States today. Liberal religion is willing to revise inherited understandings in light of reason, experience, and new knowledge. Because it is an approach to religion rather than a single denomination or religious affiliation, there is no reliable way to measure its size. But two major trends help illuminate its changing place in American religious life: the decline of historically liberal mainline churches and the growth of the religiously unaffiliated—people who do not identify with a particular religious tradition or institution. That category includes very different relationships to religion and spirituality.

Imagine our country as a small town of just one hundred people. Of those 100 people, 62 would identify as Christian—but not all in the same way:

- 23 would be evangelical Protestants
- 11 would be mainline Protestants
- 5 would belong to historically Black Protestant traditions
- 19 would be Catholic
- 2 would be Latter-day Saints (Mormons)
- 2 would belong to other Christian traditions, including Jehovah's Witnesses

Even within that majority, there is remarkable diversity. American Christianity is not a single tradition but a family with many histories, theologies, and forms of religious practice. That is only part of the story. The most significant religious change in recent decades has not primarily occurred within Christianity itself, but in the growing number of Americans who no longer claim any religious affiliation.

The rest of our little town would represent non-Christian religions as well as those with no religious affiliation:

- 2 would be Jewish
- 1 would be Muslim
- 1 would be Hindu
- 1 would be Buddhist
- 2 would belong to a wide range of other traditions, including Bahá'í, Sikh, Pagan, and Indigenous spiritualities
- 29 would be religiously unaffiliated.[3]

Where do Unitarian Universalists fit into this picture? With about 130,000 members in roughly 1,000 congregations nationwide, my chosen tradition is a very small part of the American religious landscape. In our imaginary town of 100 people, there wouldn't even be a single Unitarian Universalist.[4]

One especially helpful guide to this landscape is Ryan Burge, a political scientist and ordained American Baptist minister. Seriously—his superpower is taking the tangled complexity of American religion and turning it into clear, compelling graphs.

Liberal religion cannot be understood in isolation; its

future is being shaped by the same demographic and cultural forces transforming the entire American religious landscape. To understand the choices people are making about religious belonging, let's look more closely at those forces, beginning with the elephant in the room: **evangelical Christianity**, a strand of Protestantism shaped by personal conversion, biblical authority, and evangelism, and one that's become closely intertwined with politics in recent decades.

In the early 1970s, fewer than two in ten American adults identified as evangelical. Over the next two decades, that share grew steadily, peaking in the early 1990s at nearly three in ten Americans. Since then, the trend has gradually reversed. The decline has not been sudden or catastrophic, but it has been steady. Evangelical Christianity remains one of the largest and most influential religious movements in the United States, but it is navigating an increasingly uncertain future shaped by theological debates, political polarization, demographic change, and the broader movement of many Americans away from organized religion.[5]

If the evangelical story is marked by a dramatic surge and partial return, the **mainline Protestant** story looks very different. In the early 1970s, mainline Protestantism was the single largest religious tradition in the United States, with nearly three in ten Americans identifying with mainline denominations such as the Methodists, Episcopalians, Presbyterians, and the United Church of Christ. From that high point, however, the story is one of steady decline.

By the late 1980s, mainline Protestantism had already lost roughly one-third of its share of the population. By the early 2000s, it accounted for only about one in eight Americans. Today, just over one in ten Americans identify as mainline Protestant.

For much of the twentieth century, mainline churches exercised enormous influence on American culture, politics, education, and public life. Their decline is therefore more than a demographic story. For many people, it is personal. These are the churches that baptized them, married them, buried their loved ones, shaped their religious lives, and, for generations, stood at the center of countless communities.[6]

Looking more closely, the pattern becomes unmistakable. Over the past four decades, every one of the historic "Seven Sisters" of mainline Protestantism has experienced substantial membership decline. Their approximate membership losses, listed alphabetically by denomination, are:

- American Baptist Churches: about 24%
- Christian Church (Disciples of Christ): 74%
- Evangelical Lutheran Church in America: 45%
- Presbyterian Church (U.S.A.): 62%
- The Episcopal Church: 36%
- United Church of Christ: 57%
- United Methodist Church: roughly 40%—and now well over 50% when recent post-schism losses are fully accounted for.

Taken together, these figures tell a story not of individual denominational failure, but of a broad cultural transformation reshaping historically liberal mainline Protestantism.[7]

For readers curious about how these trends are affecting **Unitarian Universalism**, UU membership nationwide has declined by about 15% over the past decade. The scale is smaller and the trajectory gentler, but UUism is not exempt from the forces reshaping American religious life.

An equally important shift has been the rapid growth of the "**nones**": those who claim no religious affiliation. For most of the twentieth century, religiously unaffiliated Americans remained a relatively small and stable share of the population, hovering in the single digits. Then, beginning in the early 1990s, the trend changed dramatically:

- In 1993, about 9% of Americans were religiously unaffiliated.
- By 2014, that share had risen to 23%.[8]
- By 2024, 29% of American adults identify as religiously unaffiliated.[9]

The nones are now larger than either evangelicals or Catholics, the two largest individual Christian groups in this breakdown.[10]

This shift reflects a deep generational change in how Americans relate to religion. For some, it combines a continuing desire for meaning, purpose, and spirituality with a changing relationship to inherited forms of religious authority; for others, religion and spirituality have simply become less important in their lives. In that sense, American religion is being transformed as people seek belonging, wisdom, and transcendence (a sense of connection to something larger than themselves) in new ways. The search endures, even as many people leave institutions they experience as rigid, untrustworthy, or disconnected from their lived experience.

One of Burge's most helpful contributions is showing

that the religiously unaffiliated are anything but a single, uniform group. The category "nones," however, conceals almost as much as it reveals. Drawing on large national surveys, Burge and his colleagues identify four broad types of religiously unaffiliated Americans, ranging roughly from those who remain most connected to religion to those who are most distant from or critical of it.[11]

First are the **"Nones in Name Only,"** who make up about 21% of all nones. These are people who check the box "nothing in particular," but still pray, believe in God, or participate in religious or spiritual practices at least occasionally. Many continue to attend worship services from time to time and say that spirituality—and even religion—remain important in their lives. They have not so much rejected religion as declined to identify with a particular religious label.[12]

Second are the **"Spiritual But Not Religious,"** the largest group at about 36% of all nones. They are skeptical of religious institutions but remain open to spirituality, embracing practices such as meditation, nature-based spirituality, or personal rituals. They tend not to pray or attend religious services regularly, but they are far from indifferent to purpose, meaning, or transcendence. It is among people like these that many discover Unitarian Universalism and other liberal religious communities, where questions of spirituality, meaning, and purpose can be explored within an open, compassionate, and non-dogmatic framework.[13]

Third are the **"Disengaged,"** who make up roughly 33% of all nones. These are people who are done with religion. Both religion and spirituality rank very low in their lives. They do not attend religious services, rarely or never pray, and generally express little interest in spiritual questions. Instead, they are more likely to find meaning in relationships, work, creativity, nature, ethical commitments, family, and the ordinary experiences of everyday life.[14]

Finally, Burge identifies a much smaller but highly visible group he calls **"Zealous Atheists,"** comprising about 11% of all nones. These are the people most likely to criticize religion publicly, debate religious belief online, or encourage others to leave religion. Although they make up only about one in ten religiously unaffiliated Americans, they are often the most visible representatives of the nones in public conversations. As a result, they can create the mistaken impression that all religiously unaffiliated people are actively opposed to religion.[15]

Taken together, this typology shows that the rise of the nones is not a single story of people rejecting religion, but a widening range of relationships to religion, spirituality, and religious institutions.

Looking ahead, Burge identifies five broad trends—trajectories rather than firm predictions, since the long-term future remains uncertain:

First, Christianity will remain the largest religious tradition in the United States for the foreseeable future. From a demographic standpoint, America remains a majority-Christian country. Although that share has declined significantly (from about 90% of Americans in the early 1970s to roughly two-thirds today), it has by no means disappeared. The Christianity of tomorrow, however, will not look like the Christianity of the past. It will increasingly be shaped by new voices, new communities, and the growing ethnic, racial, and cultural diversity of American society.[16]

Second, religious switching is accelerating and reshaping the American religious landscape. For most of American history, the strongest predictor of a person's adult religion was the religion in which they were raised. With each successive generation, however, fewer Americans remain in their childhood religious tradition.

The largest beneficiary of this growing religious mobility has been the religiously unaffiliated. As more adults leave organized religion, more children are also being raised without a religious affiliation, creating a self-reinforcing cycle that is likely to continue reshaping the American religious landscape for years to come.[17]

Third, mainline Protestantism will continue to lose its place at the cultural center of American religion. This does not mean extinction, but rather a diminished cultural influence. Mainline churches will continue to exist and, in many places, remain strong. Yet with their share of the population much smaller and the religious landscape increasingly pluralistic, they are unlikely to shape American culture, politics, education, and public life as they did throughout much of the twentieth century.[18]

Fourth, the rapid rise of the religiously unaffiliated may itself be beginning to slow. The growth of the nones has been the single most dramatic demographic change in American religion over the past three decades. What was once a small, single-digit segment of the population now includes 29% of American adults—and well over 40% of younger Americans. But the most recent data from multiple sources suggest that this extraordinary period of growth may be entering a new phase. Rather than continuing its rapid rise, the percentage of religiously unaffiliated Americans appears to be stabilizing.[19]

Fifth—and perhaps most importantly—the future of American religion has become increasingly difficult to predict. This is less a cop-out than an act of intellectual honesty and humility. Even the best data-driven models now generate multiple plausible possibilities, depending on the patterns of religious switching in the decades ahead. In every scenario, Christianity continues to decline while the religiously unaffiliated continue to grow. But the pace, timing, and eventual balance of those trends vary considerably. What lies ahead for American religion, in other words, is not fixed. It is being shaped by the choices each new generation makes about belief, belonging, and community.[20] The numbers can tell us how patterns of religious belonging are changing. They cannot tell us what makes a community worth choosing—or why people stay.

Think for a moment about a community—religious or otherwise—you chose rather than inherited. What made you want to belong? What made you stay?

As more Americans move away from inherited religion, many are not abandoning the search for a meaningful and ethically grounded life. Some are leaving institutions they experience as rigid or coercive; others simply find religion increasingly peripheral to their lives. As inherited religious authority weakens, liberal religious communities can offer places where people seek together what is true and life-giving.

Liberal religious communities can build community without creeds, make room for depth without dogma, and offer belonging without belief tests. Across many traditions, congregations are finding new ways to nurture ethical formation, spiritual practice, and shared purpose for people no longer willing to inherit religion simply because it was handed to them.

The theologically conservative congregation of my childhood taught me to work out *my* salvation "with fear and trembling." The focus was on the next world, where each individual would spend eternity. Much of religion, as I encountered it then, was shaped by fear: fear of sin, fear of judgment, fear of hell. Liberal religion showed me another way: not to focus fearfully on my individual salvation, but to work toward *our* salvation, right here and now, in this world and on this Earth, with hope and trust, curiosity and love.

If the future of religion is becoming less inherited, more chosen, and more diverse, then the task before liberal religion is clear: to become communities worth choosing, again and again—places that put love at the center and practice truth-seeking and courage together.

In the end, the truest measure of a religious community isn't the size of its institutions. It's the kind of people it helps us become.

18

HOW TO BE A GOOD ANCESTOR

Who will have to live with what I choose?

The scientific picture is clear. Human activity is warming the planet, destabilizing ecosystems, and reshaping the conditions that have allowed human civilization to flourish. Every fraction of a degree matters, and the choices made during this century will reverberate for generations. How might we think and act differently to help co-create a more ecologically sustainable future?

My son was born in 2023. If he lives to seventy-seven, he will see the year 2100. Becoming a parent did not introduce me to concern for the future. But it gave the future a face. Perhaps someone in your life does the same—a child, a grandchild, a neighbor—anyone you care about who is young enough to live with consequences you may never see.

That shift in perspective has changed how I think about ancestry itself. I have found three books especially helpful. The first is philosopher Roman Krznaric's *The Good Ancestor: A Radical Prescription for Long-Term Thinking*. One of its most compelling ideas is that ancestry points in two directions: we honor those who came before us, but for generations yet unborn, we are someone's ancestors.

Those of us alive today inherit the work of those who came before us. Our lives are richer and easier because of generations of inventors, activists, artists, caregivers, and others whose names we'll never know. As the old saying goes,

> We drink from wells we did not dig.
> We sit in the shade of trees we did not plant.
> We profit from the sacrifice of people we did not know.[1]

There's no way to pay that back. But we can pay it forward.

Thinking like a good ancestor requires expanding the horizon of our moral concern. Instead of asking only what kind of life we want to live, we begin asking what kind of world we want to leave behind. Our attention shifts from ourselves alone to generations ahead that we will never meet.

One distinctive contribution of liberal religion has been to redirect attention from the next world to this one. As a Unitarian Universalist saying puts it, instead of fearing hell in the next world, our task is to "love the hell out of this world."[2]

The American pragmatist philosopher Richard Rorty captured this shift beautifully when he suggested that our willingness to sacrifice for future reward can be redirected "from one's hope for paradise to one's hopes for one's grandchildren."[3]

This idea carried particular meaning in his own family. Rorty's grandfather was Walter Rauschenbusch, one of the founders of the Social Gospel movement, which urged Christians to focus less on individual salvation and more on building a more just society. Rorty could never repay the intellectual debt he owed a grandfather who died more than a decade before he was born. But he could honor that inheritance by carrying its deepest aspirations forward, reshaping them for a new generation.

One way to appreciate this inheritance is to consider the mathematics of ancestry. Tracing a family tree backward, the number of ancestral places doubles with each generation:

- 2 parents
- 4 grandparents
- 8 great-grandparents
- 16 great-great-grandparents
- 32 ancestors (5 generations back)
- 64 ancestors (6 generations back)
- 128 ancestors (7 generations back)
- 256 ancestors (8 generations back)
- 512 ancestors (9 generations back)
- 1,024 ancestors (10 generations back)
- 2,048 ancestors (11 generations back)
- 4,096 ancestors (12 generations back)[4]

Behind those places on your family tree were thousands of lives. How many hard winters, buried children, love affairs, small mercies, and stubborn hopes for a future they'd never see had to come before you could exist in this present moment? One day, someone may look back on us as their ancestors. What kind of inheritance will we leave behind?[5]

One powerful example of paying it forward comes from physician and medical researcher Jonas Salk. In 1955, Salk and his team developed the first effective vaccine against polio, a disease that had long terrified parents and children around the world. Rather than patenting the vaccine, Salk insisted that it belonged to humanity. When asked who owned the patent, he replied, "Could you patent the sun?"

Reflecting on what sustained him through years of painstaking research, Salk said that he was driven by the desire "to be of some help to humankind." Later in life, he repeatedly returned to a question that lies at the heart of this chapter: "Are we being good ancestors?"[6]

Krznaric argues that becoming a better ancestor requires resisting six habits of short-term thinking while developing six corresponding ways of seeing the world more expansively:

- **Tyranny of the Clock → Deep-Time Humility**: From focusing only on today's demands to remembering that we are but an eye blink in cosmic time.
- **Digital Distraction → Legacy Mindset**: From having our attention hijacked by technology to asking how we hope to be remembered.

- **Political Presentism → Intergenerational Justice:** From obsessing over the next election to considering the well-being of the seventh generation.
- **Speculative Capitalism → Cathedral Thinking:** From pursuing short-term gains to undertaking projects whose benefits may outlast our own lifetimes.
- **Networked Uncertainty → Holistic Forecasting:** From merely reacting to crises to imagining multiple possible futures.
- **Perpetual Progress → Transcendent Goal:** From assuming endless economic growth to pursuing "one-planet thriving."[7]

Each of these shifts offers a different way of stretching our moral imagination beyond the present moment. Let's look more closely at each one.

First, Deep-Time Humility: learning to see ourselves within the vast sweep of evolutionary and cosmic history. Around the world, governments, Indigenous communities, scientists, and conservation organizations are already beginning to think on longer timescales, for example, by protecting biodiversity through long-term seed banks and establishing protected lands intended to endure for generations.[8] Can we make that kind of long-term thinking more common? What long-term commitments are we making today? And how can we build institutions capable of sustaining them beyond the shifting winds of election cycles and changing political priorities?[9]

Second, Legacy Mindset: How do we hope to be remembered?[10] As Salk's example suggests, one of the deepest sources of meaning comes from shifting our attention away from accumulating more for ourselves and toward leaving behind a life that continues to bless others long after we are gone.

Third, Intergenerational Justice: Groucho Marx was joking when he quipped, "Why should I care about future generations? What have they ever done for me?" But many Indigenous traditions have long embraced the opposite perspective: we are not owners of this planet but temporary stewards, holding it in trust for those who will come after us. As a widely quoted saying puts it: "We do not inherit the Earth from our ancestors; we borrow it from our children."[11]

Greta Thunberg gave this principle a sharp contemporary expression when she told world leaders, "You have stolen my dreams and my childhood with your empty words.... The eyes of all future generations are upon you. And if you choose to fail us, I say: we will never forgive you."[12] Her challenge is that our concern for future generations must be measured by the world we leave them. Limiting global warming as much as possible will require acting on behalf of people we will never meet—and responding to the crisis with action equal to its scale.[13]

Fourth, Cathedral Thinking: Medieval cathedrals were so ambitious that the people who began building them often knew they would never live to see them completed. In some cases, several generations of builders would come and go before the work was finished. Cathedral thinking means taking on projects whose fruits may be enjoyed only by forthcoming generations.

Fifth, Holistic Forecasting: Wise long-term planning

requires looking beyond any single measure of success and considering how today's choices affect interconnected systems over time. For too long, our economic system has treated short-term financial profit as the only bottom line. But what if we measured success more broadly? What if we also asked whether our choices strengthened communities, restored ecosystems, expanded opportunity, and left the next generations with a more livable world?

One promising example is the growing movement of Certified B Corporations (or B Corps), which are evaluated for their social and environmental impact, accountability, and treatment of stakeholders. Patagonia, for example, is a certified B Corp whose business practices are evaluated for their effects on workers, communities, customers, and the environment as well as for how the company is governed. Their approach reflects what is often called the triple bottom line: people, planet, and profit. Financial sustainability still matters, but it is understood as one goal among several rather than the only one.[14]

In much of today's economy, success is still measured primarily by quarterly earnings and shareholder returns. B Corps point toward a different possibility: an economy in which businesses are expected to create value not only for investors, but also for workers, communities, the living world, and future generations. Imagine how different our economy might look if every major corporation had to answer for its effects on workers, communities, and ecosystems as well as its bottom line. Like every human institution, our economic system is not fixed. It reflects choices—and different choices remain possible.

Sixth, Transcendent Goal: Instead of assuming that

endless economic growth is either possible or desirable, Krznaric asks what human flourishing might look like within the ecological limits of a single planet. The question becomes not simply how much more we can produce and consume, but what kind of flourishing can endure.

The second book is *Zen and the Art of Saving the Planet* by Thích Nhất Hạnh, one of the founders of Engaged Buddhism. He taught that meditation is not only about waking up as individuals, but also about developing the wisdom and compassion needed to help awaken the world.

Especially for those who feel exhausted or overwhelmed by the scale of the climate crisis, Nhất Hạnh taught his students to cultivate a wise balance between contemplation and action, related to the Buddhist teaching of the Middle Way:

- Sometimes he would say, “Don’t just do something, sit there.”[15] Action divorced from wisdom or compassion can easily undermine the very future we hope to protect. In such moments, the most appropriate response begins by slowing down, becoming fully present, and allowing clarity and compassion to guide our actions.
- Other times, when suffering demanded an immediate response, Nhất Hạnh urged his students to postpone even basic comforts in order to help others.
- Still other times, when he saw that they had become exhausted, he quietly prepared food for them himself. He understood that healing the

> world also requires caring for those who seek to heal it.[16]

The key is learning to discern which moment we are in. Is this a time to pause, listen, and renew your spirit, or a time to step forward with courage? What is the next right action at the intersection of wisdom and compassion? What are future generations—and the Earth itself—asking of you?

The third book poses a more uncomfortable challenge. In *How to Blow Up a Pipeline*, Andreas Malm, a Swedish scholar of human ecology and a climate activist, wrestles with whether the urgency of the climate crisis calls for more disruptive forms of civil disobedience than many environmental movements have been willing to embrace.[17]

Malm is not advocating violence against people or other life. Rather, he asks whether carefully targeted acts of property sabotage might, under certain circumstances, be ethically justified as a form of climate resistance.[18]

He also challenges the environmental movement to ask whether marches, rallies, speeches, and other forms of symbolic protest are sufficient to confront a crisis of this magnitude. Such activism risks growing into little more than "ritualized wishful thinking" if it fails to disrupt business as usual.[19]

I have no plans to blow up pipelines, but reading Malm's book forced me to ask what this moment actually requires of us. Future generations are unlikely to judge us by whether we approved of every tactic. They will judge us by whether we recognized the emergency and acted while there was still time.

For most of us, the choices will be less dramatic. We

belong to institutions making decisions whose consequences will outlast us: congregations, workplaces, schools, and local governments. Good ancestry can begin by choosing one of those places, asking what decisions are being made there today that people decades from now will have to live with, and finding one concrete way to help shape them. What's the next right step within your sphere of influence? And after that, what's the next one?

We are not waiting for a new heaven and a new Earth. This is our paradise planet. This is our blue boat home.

19

BECAUSE I'M IN LOVE

What is worth doing without knowing how it ends?

Discouragement has become a major obstacle to becoming a good ancestor. Humanity now knows more than ever about climate change, biodiversity loss, mass extinction, and the ecological crises already upon us. For many people, the future increasingly feels less like a promise than a threatening forecast. It becomes tempting to ask not only, "What should we do?" but "Can anything we do still matter?"

In February 2024, *The New York Times* ran a headline that read, "Scientists Are Freaking Out About Ocean Temperatures."[1] Among my least favorite things is sober, data-driven scientists freaking out about anything. When cautious people start to panic, it's worth paying attention. Here is how one professor at the University of Miami described what he was seeing: "The North Atlantic has been record-breakingly warm for almost a year now. It's just astonishing. Like, it doesn't seem real."

A climate scientist at the University of New South Wales expressed similar concern: "The temperature's just going off the charts. It's like an omen of the future." At the time, these warnings sounded almost unbelievable. Looking back after Hurricanes Helene and Milton, they now seem chillingly prescient.

The data themselves were equally striking. Since 1979, daily sea-surface temperatures had fluctuated within a relatively predictable range. In 2023, they rose noticeably above previous records. Then, in 2024, they climbed dramatically higher still. Scientists warned that unusually warm Atlantic waters could fuel an exceptionally intense hurricane season.[2]

Seven months later, those warnings became reality. Hurricane Helene devastated western North Carolina and many surrounding communities. Shortly afterward, Hurricane Milton carved a destructive path across Florida, spawning deadly tornadoes along the way.

For nearly four decades, I've spent part of almost every year in the Black Mountain and Asheville region of North Carolina. Seeing the photos and videos was personally devastating—though my grief is nothing next to what the people living there endured. Historic Biltmore Village went underwater. The River Arts District: total devastation. Entire homes floated away across Appalachia. One of the destroyed roads runs past the corner of Lake Eden Road and Old U.S. 70 (a mile and a half from Camp Rockmont, where I spent every summer from ages ten to twenty-one) and is the only road to a high school that serves more than six hundred students.

According to *The Washington Post*, Asheville had long been considered a climate haven, a place many believed would be relatively insulated from the worst effects of global warming because of its cooler temperatures and inland location. Some even worried that climate migration would eventually overcrowd the city. Then the floods came.[3]

This continuing string of record-breaking temperatures confirms what scientists have warned for decades: climate change is already reshaping lives and landscapes. The growing crisis demands an equally ambitious response. In truth, we needed global action decades ago.

In 1970, on the first Earth Day, we already had compelling evidence that we could no longer pretend we were separate from the living world that sustains us. The obstacles to action have rarely been mainly technical. More often they are human: greed, organized disinformation, and willful denial. Too many of us have spent decades arguing about whether the crisis is real while the crisis itself proceeded on its own schedule, indifferent to our arguments. As Wendell Berry once put it, "Whether we and our politicians know it or not, Nature is party to all our deals and decisions, and she has more votes, a longer memory, and a sterner sense of justice than we do."[4]

Hannah Ritchie's *Not the End of the World: How We Can Be the First Generation to Build a Sustainable Planet* offers one of the clearest and most evidence-based explorations of what it would take to help avert ecological collapse and preserve a livable planet for generations to come. An environmental data scientist at the University of Oxford, Ritchie provides a concise, accessible, and ultimately optimistic guide to the changes that are both necessary and still possible. She examines seven areas where large-scale action is most urgently needed: air pollution, climate change, deforestation, food systems, biodiversity loss, ocean plastics, and overfishing.

Her book shows that a more sustainable future remains within reach. But practical solutions are only part of the challenge. We also need the inner resources to live wisely, courageously, and compassionately in a world that often feels uncertain. For that, we can also turn to the poets.

The title poem of Jorie Graham's *[To] the Last [Be] Human* has stayed with me ever since I first read it. I do not love every poem in the collection, but I am moved by the vision that animates it. Graham frames the book against the accelerating climate crisis, noting in her introduction that its poems were written as atmospheric carbon dioxide rose from 373 to 414 parts per million between 2002 and 2020. Her collection has been described as "an extraordinary lyric record of those eighteen calamitous years...rife with hope and raw with loss."[5]

The title poem wrestles with what it means to seek guidance and beauty amid accelerating ecological collapse. It asks what it means to call anything "okay" while the future itself feels increasingly uncertain. Reading it against the changes of recent years, I cannot help but feel that its urgency has only deepened. More has disappeared. More has changed.

I refuse to believe that the appropriate response is resignation. As a parent, I cannot give up on helping to ensure that my son—and all who come after him—inherit a planet that remains beautiful, alive, and capable of sustaining the rich diversity of life, where future generations can continue to discover how to belong.

Is that merely wishful thinking? Are we really capable of the changes needed to move from accelerating ecosystem collapse toward cultivating a thriving, sustainable planet for generations to come? Two quotations have helped me hold together realism and possibility.

The first comes from Fredric Jameson, who is perhaps best known for the observation that "Someone once said that it is easier to imagine the end of the world than to imagine the end of capitalism."[6]

The second quotation comes from Ursula K. Le Guin. In her acceptance speech for the National Book Foundation Medal for Distinguished Contribution to American Letters, she observed: "We live in capitalism, its power seems inescapable—but then, so did the divine right of kings. Any human power can be resisted and changed by human beings."[7]

I am not invoking these quotations to argue that capitalism must end altogether. But I do think our economic systems will need to change significantly if we are to live within the limits of a finite planet. The broader point is this: human arrangements that seem permanent can change.

We make a mistake when we assume that the future can only resemble the past or the present. History is filled with changes that once seemed impossible until they happened. Human beings can adapt to changing circumstances, imagine new possibilities, and sometimes create them.

Imagining a different future is one vital step toward making it possible. We cannot control what comes next, but our choices help shape it—and perhaps help bend what the nineteenth-century Unitarian minister Theodore Parker described as "the arc of the moral universe" toward justice, a phrase later made famous by Martin Luther King Jr.[8]

In that spirit, I appreciate the activist Derrick Jensen's reflection on what has sustained him in the work of bending that arc toward climate justice:

> Why am I an activist? Because I'm in love. With salmon, with trees outside my window, with baby lampreys living in sandy stream bottoms, with slender salamanders crawling through the duff. And if you love, you act to defend your beloved. Of course results matter to you, but they don't determine whether or not you make the effort. You don't simply hope your beloved survives and thrives. You do what it takes. If my love doesn't cause me to protect those I love, it's not love.[9]

Too often, environmentalism is framed primarily in terms of fear, guilt, sacrifice, or obligation. Those motivations may sometimes spur us to action, but they rarely sustain us over a lifetime.

Love can.

Whether we speak of love for what is sacred, love for neighbor, love for future generations, or love for the Earth itself, love has always been one of humanity's deepest sources of moral courage. If we love this Earth—its forests and rivers, its birds and insects, its human and more-than-human neighbors—then caring for it stops being a burden and becomes devotion.

If love is to become more than a feeling, it must be translated into effective collective action. One organization that has helped me think about how to do that is ecoAmerica, a national nonprofit that seeks to "build institutional leadership, public support, and political resolve for climate solutions in the United States." Among its most helpful resources are the Climate Action Guides, which offer practical, evidence-based advice for communicating about climate change in ways that are both compassionate and effective. Drawing on their work, here are five principles that I have found especially helpful.[10]

If you've ever had a frustrating conversation about climate change, you're not alone. No communication strategy will persuade everyone. But ecoAmerica's research highlights five practices that help make these conversations constructive and effective.

First, begin with people rather than arguments. Listen before you speak. Start with the concerns and values your neighbors already hold (love for family, care for their community, health, fairness, or economic security) and connect climate action to those shared commitments.

Second, make climate change concrete. Talk about changes your fellow citizens can see in their own communities, whether shifting seasons, stronger storms, prolonged droughts, or local flooding. "Stories sell, numbers numb."

Third, emphasize solutions rather than dwelling only on sacrifice. Many people remain unaware of how quickly renewable energy, battery storage, electric vehicles, and other technologies are advancing. Climate action is not simply about avoiding catastrophe; it is also about cleaner air, healthier communities, new jobs, innovation, and long-term economic resilience.

Fourth, inspire hope and a sense of agency. "So ditch

the doom and gloom." Fear can capture attention, but hope is more likely to sustain action. Help members of your community recognize that their choices matter—not because any one person can solve the climate crisis, but because action spreads. When a few people start something concrete in a congregation or neighborhood, others see that change is possible and are more likely to join in.

Finally, make the conversation actionable. Avoid getting trapped in endless arguments over misinformation. Keep the larger picture in view, speak with respect, offer practical options, and close with one clear action people can take: sign up for a local climate group, contact an elected official, or support a concrete change in your community.

Notice that all five begin not with winning an argument but with cultivating a relationship. Lasting change rarely begins with being right. More often, it begins when people feel seen and respected and can recognize themselves as part of a larger story.

Let's return one final time to Jorie Graham. In another poem from *[To] the Last [Be] Human*, titled "Poem," Graham imagines the Earth itself speaking to us. Again and again, the Earth itself pleads not to be forgotten.

That refrain is worth carrying with us. Remember the forests. Remember the oceans. Remember the atmosphere that makes every breath possible. Remember the countless forms of life with whom we share this planet. Remember those who came before us and those who will come after us. Above all, remember that we are Earthlings. We do not just live on this planet. We belong to a living world.

20

APOCALOPTIMISM

What can I build while seeing clearly what is broken?

Hope is not the same thing as optimism. It does not require believing that history inevitably bends toward justice or that progress, once achieved, cannot be reversed. Hard-won gains can be lost.

The opening months of President Trump's second administration served as a sobering reminder of that reality. Environmental policy offered one stark example: the administration prioritized expanding fossil fuel production, weakening environmental protections, and scaling back federal efforts to address climate change.[1] Douglas Brinkley, a widely respected presidential historian at Rice University, said it's not an exaggeration to call Trump "the worst president for the environment in American history."[2]

These developments also bring to mind a powerful book by Gus Speth, cofounder of the Natural Resources Defense Council and former dean of Yale's School of Forestry and Environmental Studies, titled *They Knew: The US Federal Government's Fifty-Year Role in Causing the Climate Crisis*. Speth has said that he once believed the greatest environmental challenges were problems such as biodiversity loss and climate change themselves. Over time, though, he decided the top problems weren't scientific at all. They were selfishness, greed, and apathy—and dealing with those, he said, would take a spiritual and cultural transformation.[3]

If the ecological crisis also has a spiritual dimension, what kind of spirituality is equal to this moment? One answer I have found comes from a concept I learned from my meditation teachers, Vince and Emily Horn. They use the term apocaloptimism to describe a way of meeting our ecological moment with both clear eyes and an open heart.

Apocaloptimism combines "apocalypse" and "optimism." Apocalypse comes from an ancient Greek word meaning "uncovering" or "revelation." An apocaloptimist faces how dire things are while remaining alert to possibilities that may become visible only in a crisis. The concept joins clear-eyed realism with confidence in our capacity for resilience.

A saying associated with the spiritual teacher Ram Dass captures this spirit: "If enlightened, I will show up with open heart and willing hands. If apocalypse, I will show up with open heart and willing hands."[4]

What might those willing hands build? Ezra Klein and Derek Thompson's *Abundance* offers one provocative answer.

The book has generated spirited debate among liberals, and I understand many of the concerns critics have raised. Some of the most valuable books are those that challenge us without persuading us completely. *Abundance* was one of those books for me.

As a longtime fan of Klein's work, I preordered it months before it was published, expecting a thoughtful book about policy and housing and energy. But after just a few pages I found myself thinking, "There's an Easter sermon in this book!"

Klein and Thompson's central argument is that resisting what is broken is only half the work; we must also build what is needed. Whether or not you agree with every proposal, *Abundance* is a hopeful reminder that the future is still open: that we can still build something better than what we have.

For a moment, imagine what life on this planet might be like twenty-five years from now if we found a way to overcome the selfishness, greed, and apathy that so often distort our politics. Don't analyze it yet. Just let yourself picture an ordinary day in that world.

Picture waking up in the year 2050 in a home powered almost entirely by clean energy. Solar panels quietly harvest the morning sun. Wind, geothermal energy, advanced nuclear technologies, and other low-carbon sources provide abundant electricity that is both affordable and nearly carbon-free. People living then might find it astonishing that we once powered our civilization by digging fossil fuels out of the ground and filling the atmosphere with pollution that destabilized the climate. In the future, as Klein and Thompson put it, that may well seem barbaric.

Klein and Thompson go on to imagine many other ways such a future could be more prosperous, more equitable, and more environmentally sustainable. Whether every prediction pans out matters less than this: the book asks us to imagine a future worth building, and then go build it.

One of the book's most compelling arguments is its discussion of what Klein and Thompson call the Green Dilemma. For decades, environmental movements have become exceptionally skilled at preventing environmental harm by saying no to projects that threaten ecosystems, pollute rivers, destroy forests, or drive species toward extinction. Those efforts have protected countless places and prevented immeasurable harm.

But if we are serious about realizing an abundant, low-carbon future, we also need to become serious about building. Achieving anything like the better world Klein and Thompson envision would require constructing clean-energy infrastructure on an enormous scale and at extraordinary speed. As they point out, this would involve "multiple national-scale infrastructure initiatives of magnitudes never before processed through existing... environmental law standards and procedures."[5]

Many of our environmental laws were enacted to slow, stop, or redirect development that threatened ecosystems, clean air, clean water, or endangered species. Those protections remain deeply important. Yet some of the very procedures designed to protect nature can also delay or obstruct the rapid deployment of the clean-energy infrastructure needed to address climate change itself.[6] That creates real tradeoffs. Building transmission lines, renewable-energy projects, and other infrastructure at the scale and speed the climate crisis requires will sometimes mean accepting local environmental impacts and streamlining review processes, while still protecting against serious harm. The challenge is to make those tradeoffs wisely rather than pretending they do not exist.

Achieving a Green New Deal will require wise and skillful people in charge of governance. Right now we have the opposite. The Trump administration's Department of Government Efficiency (DOGE) was a prime example of how to move fast and break things with maximum cruelty and complete disregard for people and the Earth. Speed, by itself, is not a virtue. The best way forward lies in knowing what deserves to be accelerated and what deserves to be protected. Rather than dismantling the institutions needed to meet shared challenges, we need to build the capacity required for a more sustainable, just, and abundant future.

Abundance challenged me to think more creatively and more constructively about what it would take to build a flourishing, low-carbon tomorrow. Reading, however, is only the beginning. Every community needs people willing to help shape that better world. Whether your passion is climate action, democracy, immigration, racial justice, LGBTQ+ equality, conservation, education, or some other form of public service, where are you being called to help build the society that coming generations deserve?

A poem titled "Feb. 2" by Margo Stebbing, written at the end of the second week of Trump's second presidential term, traces a movement from despair toward renewed resolve. It begins with a speaker exhausted by years of warning about environmental destruction and democratic decline. Then an encounter with a flock of Canada geese becomes an unexpected turning point. Standing in their quiet presence, the speaker remembers that grief need not become surrender and that, as long as we are still alive, our work is unfinished.

That's the sort of resurrection I can begin to believe in. That's apocaloptimism: whatever comes, may we show up with an open heart and willing hands.

AFTERWORD

Although I conceived this book only recently, its chapters began as sermons preached over more than a decade. The earliest sermon dates to 2013, my first year as minister of the Unitarian Universalist Congregation of Frederick, Maryland; the most recent was preached in 2026. At the time, each sermon felt like its own conversation, prompted by a particular book, season, or event. Only in retrospect did I begin to see how they might become part of a single book.

Writing this book has been primarily an act of discovery. As the chapters came together, I realized that what united them was not a single subject but a way of seeing. Ecology, evolution, religion, attention, democracy, hope, and climate change all turned out to be different paths toward the same question: What does it mean to belong to a living world?

Perhaps that is fitting. If evolution teaches us anything, it is that growth rarely follows a straight line. It proceeds through small adaptations, false starts, and dead ends. Ideas, people, and religious and spiritual traditions evolve much the same way.

The book has ranged from the birth of the universe to the emergence of consciousness, from the inner lives of animals to the hidden intelligence of forests and fungi, from meditation to democracy, climate change, and the generations yet to come. Though these subjects may seem scattered, I've come to believe they all ask the same thing of us: to become more fully Earthlings.

To be an Earthling is to remember that we belong not only to our families, nations, religions, or political parties, but to the community of life itself. We are neither the center nor the purpose of the universe. We are participants in a story billions of years older than ourselves—and one that will continue long after we are gone.

That realization is both humbling and liberating. Meaning need not be handed down from outside the universe. It comes from the communities we build, the beauty we make, and the suffering we help relieve.

The spirituality that has emerged across these chapters is not about escaping this world, but about learning to dwell within it more wisely and compassionately. Liberal religion is strongest when it asks us to hold our convictions firmly enough to act on them, yet lightly enough that reality can keep teaching us. It also calls us to widen our circle of moral concern: to include those who differ from us, those who will inherit the future, and the other forms of life that share this planet with us.

None of us will live this vision perfectly. I certainly won't. There will be days when we give in to fear, cynicism, or despair. There will be times when the problems before us seem impossibly large and our own contribution painfully small. When those moments come, I'd ask you—and myself—to remember two things.

First, we belong. None of us stands apart or acts alone. Our lives depend upon other people, other creatures, the living world around us, and generations whose work made our own lives possible. Every movement for justice, every work of art, and every lasting work of repair builds on the efforts of others. The question is not whether that larger story includes us, but what part we will choose to play in it.

Second, what we do matters. Not everything is in our hands, and no outcome is guaranteed. But because we belong to one another and to the living world, our choices do not end with us. They shape our relationships, our communities, and what others inherit. We may never know the full consequences of what we do. Our part is still ours to play.

We have inherited both great gifts and hard challenges. May we prove worthy of both.

And may we become, ever more fully, Earthlings—learning together how to belong to a living world.

AUTHOR'S NOTE

I've been an ordained minister since 2003. During my first seven years of ministry, I served as an associate pastor and delivered sermons only occasionally. Since becoming the lead minister of a congregation in 2010, however, I have preached more than thirty-five sermons a year on average.

During my twenty-third year of full-time ministry, I realized that my years of weekly preaching, together with my earlier ministry, had added up to more than 600 sermons. At an average length of roughly 2,400 words, that amounts to well over 1.4 million words: the equivalent of nearly 20 typical nonfiction books. The difference is that those "books" are scattered across hundreds of sermons, each written for a particular congregation at a particular moment in time.

The easiest approach would have been to publish thematic sermon collections. But I found myself drawn to a more challenging—and, I hope, more rewarding—project:

returning to those sermons and reshaping them into something new. My aim has been to make the whole greater than the sum of its parts.

I became especially interested in what might happen if sermons originally preached weeks, months, or years apart were placed in direct conversation with one another. Would ideas that had seemed unrelated turn out to belong together? Would patterns emerge that I had not recognized while writing them? Bringing those sermons alongside one another became a kind of experiment: a way of discovering what I had been thinking about all along.

My homiletics professor in seminary used to say, "A sermon should be about something sacred...and about twenty minutes." I've tried to honor that advice for more than two decades. But some questions refuse to fit inside twenty minutes. What does it mean to be an Earthling? How should religion change in light of evolution, ecology, and modern science? How do we hold on to hope in an age of climate change and democratic uncertainty? Those are questions that need a different kind of conversation: one with room to wander and wonder, and to follow connections that would never fit into a single sermon. A book makes that kind of conversation possible.

To write this book, I revisited every sermon. I removed dated references, filled in gaps, reorganized ideas, and discovered connections that I had not seen when each sermon stood on its own. In the process, I also came to understand my own work more clearly. What I had assumed were hundreds of separate sermons now strikes me as one long conversation.

This book gathers one part of that conversation into a form that can stand on its own. I hope it is only the beginning.

ACKNOWLEDGMENTS

No book is ever written alone. Although my name appears on the cover, this one is the product of countless conversations, friendships, teachers, books, and communities that have shaped me over many years.

This book owes an enormous debt to the members of the Unitarian Universalist Congregation of Frederick. Many of these chapters began as sermons first preached in that beloved community, and over the years your questions, stories, encouragement, disagreements, laughter, and willingness to explore unfamiliar ideas have shaped both this book and the person who wrote it. Thank you for giving me the extraordinary privilege of thinking aloud with you week after week. Whatever wisdom these pages contain, much of it grew from our life together.

I am especially grateful to the late Jim Edwards, whose friendship and mentorship first introduced me to the riches of liberal religion. Long before I imagined writing this book, he helped me discover a way of being religious that welcomed curiosity, honored evidence, and trusted that truth has nothing to fear from honest questions. This book is, in many ways, one small attempt to pass that gift along.

I am also grateful to David Rutledge, my undergraduate advisor in religion. His book *Humans and the Earth: Toward a*

Personal Ecology was an early introduction to ways of thinking about the relationship between spirituality and the living world.

I owe an immeasurable debt to the countless authors whose ideas have accompanied me over the years. Although only a fraction of them appear in these pages, they have all helped shape the way I think about science, religion, philosophy, ecology, democracy, meditation, and what it means to live a meaningful life. Reading has always been one of the great joys of my life, and this book is a long conversation with teachers I have met both in person and through their books.

I am especially grateful to my early readers, whose careful attention, candid questions, corrections, and encouragement made this a stronger book: Kathryn Adams, Steve Berté, Lisa Bromfield, Bob Clegg, Bob Cornet, Barbara Kershner Daniel, Bruce Fry, Nita and Spencer Hines, Polly Matzinger, Pam McDonald, Penny McDougal, Meg Menke, Kathy Myers, Jude Oyler, Nancy Pace, Tom Pollack, Polly Zahn Schmitt, Andy Shelton, Mike Vandre, and Richard L. Warren.

I am grateful to Vince Fakhoury Horn and Emily Horn, whose teaching and friendship have deepened my experience of meditation, dharma, and contemplative practice.

Finally, my deepest thanks belong to my family.

To my wife, Magin LaSov Gregg, thank you for your love, patience, and encouragement. Watching you pursue your own book with such dedication has been a constant source of inspiration.

And to my son, Jaylen. You will be much older before you read these words. Much of this book was written while watching you discover the world with fresh eyes. Your curiosity, delight, and wonder continually reminded me why the future matters. More than anyone else, you renewed my determination to become the kind of ancestor this book invites us all to be.

May this book, in some small way, help build a future worthy of you—and of all who will inherit this beautiful, fragile Earth.

FURTHER READING

"Salvation by bibliography."

The books listed below most deeply formed my thinking while writing this book. They are not exhaustive bibliographies but invitations to continue the conversation.

Introduction

Gary Dorrien. *The Spirit of American Liberal Theology*.

To Dwell Poetically

James C. Edwards. *The Plain Sense of Things*.

Insignificant Animals, Dissatisfied Gods

Yuval Noah Harari. *Sapiens*.

Your Dog Really Does Love You

Carl Safina. *Beyond Words*.
Peter Godfrey-Smith. *Other Minds*.

Look and See
Jim Holt. *Why Does the World Exist?*

What Doesn't Go Away
Steven Weinberg. *To Explain the World.*

The Sun Doesn't Rise
David Christian. *Origin Story.*

Stop Being One Person
Greta Thunberg, ed. *The Climate Book.*

Choosing Earth
Kelly Weinersmith and Zach Weinersmith. *A City on Mars.*

Hidden in Plain Sight
Peter Wohlleben. *The Hidden Life of Trees.*

Relationships All the Way Down
Merlin Sheldrake. *Entangled Life.*

Same Room, Different Worlds
Ed Yong. *An Immense World.*

Before Allegiance
Robin Wall Kimmerer. *Braiding Sweetgrass.*

We Become What We Attend To
Gloria Mark. *Attention Span.*
Jenny Odell. *How to Do Nothing.*

Never Say Never, Never Say Always

Alan Levinovitz. *Natural.*

Two Weeks on a Picnic Table

Cal Newport. *Slow Productivity.*

Tricia Hersey. *Rest Is Resistance.*

Coming to Our Senses

Kristin Neff and Christopher Germer. *The Mindful Self-Compassion Workbook.*

Reginald A. Ray. *Somatic Descent.*

Sonya Renee Taylor. *The Body Is Not an Apology*, 2nd ed.

Communities Worth Choosing

Ryan P. Burge. *The American Religious Landscape.*

How to Be a Good Ancestor

Roman Krznaric. *The Good Ancestor.*

Thích Nhất Hạnh. *Zen and the Art of Saving the Planet.*

Andreas Malm. *How to Blow Up a Pipeline.*

Because I'm in Love

Brian D. McLaren. *Life After Doom.*

Apocaloptimism

Ezra Klein and Derek Thompson. *Abundance.*

NOTES

Epigraph

1. Thích Nhất Hạnh, "Interrelationship," in *Call Me by My True Names: The Collected Poems of Thich Nhat Hanh* (Berkeley, CA: Parallax Press, 2005), 47.

Introduction

1. For more about Unitarian Universalism, visit https://www.uua.org.
2. Gary Dorrien, *The Making of American Liberal Theology, Volume I: Imagining Progressive Religion, 1805–1900* (Louisville, KY: Westminster John Knox Press, 2001), 1.

1. To Dwell Poetically

1. James Edwards, "A Skeptical Perspective," in *Furman Studies: Furman University's Religious Identity* (Greenville, SC: Furman University, 1997), 160–161.
2. James Edwards, "A Skeptical Perspective," 162.
3. Edwards's phrase "to dwell poetically on the earth as a mortal" echoes a line by the German poet Friedrich Hölderlin: "poetically man dwells." A century later, Martin Heidegger built an essay around it, asking what dwelling means when it is more than occupying a house.
4. For a contemporary theological development of *poiesis*, see John D. Caputo, *A Radical God: Theopoetics, Phenomenology, and the Future of Religion* (New York: Columbia University Press, 2026). Although his explicitly theological and deconstructive project differs from Edwards's, both understand religion less as assent to fixed doctrines than as a creative way of shaping and inhabiting a life—and of learning to belong within reality as it is.
5. James C. Edwards, *The Plain Sense of Things: The Fate of Religion in an Age of Normal Nihilism* (University Park: Pennsylvania State University Press, 1997), ix.
6. Edwards, *The Plain Sense of Things*, 223.
7. Annie Dillard, *The Writing Life* (New York: Harper & Row, 1989), 32.

8. Edwards, *The Plain Sense of Things*, 210.
9. Edwards, *The Plain Sense of Things*, 152.
10. James C. Edwards, "The Common Ground of Athens and Jerusalem," in *What Really Matters?: A Collection of Lectures from the L. D. Johnson Series*, ed. Thomas O. Buford (Macon, GA: Mercer University Press, 2001).
11. Edwards, *The Plain Sense of Things*, 240.

2. Insignificant Animals, Dissatisfied Gods

1. Yuval Noah Harari, *Sapiens: A Brief History of Humankind*, trans. Yuval Noah Harari, with the help of John Purcell and Haim Watzman (New York: Harper, 2015), 4.
2. Harari, *Sapiens*, 5–8.
3. Harari, *Sapiens*, 11–12.
4. Harari, *Sapiens*, 163.
5. Harari, *Sapiens*, 120.
6. Harari, *Sapiens*, 120.
7. Harari, *Sapiens*, 108–110.
8. Harari, *Sapiens*, 110–111.
9. Elizabeth Kolbert, *The Sixth Extinction: An Unnatural History* (New York: Henry Holt and Company, 2014), 13. The five major mass extinctions generally recognized by paleontologists are the end-Ordovician extinction (about 444 million years ago); the Late Devonian extinctions (roughly 372–359 million years ago); the end-Permian extinction (about 252 million years ago, the largest known mass extinction); the end-Triassic extinction (about 201 million years ago); and the end-Cretaceous extinction (66 million years ago), which eliminated the non-avian dinosaurs along with many other species. Each involved the loss of a substantial proportion of Earth's species and a major reorganization of ecosystems.
10. Harari, *Sapiens*, 350–351.
11. Harari, *Sapiens*, 416.

3. Your Dog Really Does Love You

1. Carl Safina, *Beyond Words: What Animals Think and Feel* (New York: Henry Holt and Company, 2015), 20.
2. Psalm 8.
3. Kolbert, *The Sixth Extinction*, 183.
4. Safina, *Beyond Words*, 79–80.

5. Safina, *Beyond Words*, 80.
6. Chen and Li's figure of ~1.2% reflects nucleotide substitutions in aligned, non-repetitive DNA. Counts that also include insertions, deletions, and repetitive sequences yield a larger figure, often cited as around 4–5%. These figures are not contradictory; they measure genomic difference in different ways. The deep similarity between the two genomes, and the common ancestry it reflects, is not in dispute. See Feng-Chi Chen and Wen-Hsiung Li, "Genomic Divergences between Humans and Other Hominoids and the Effective Population Size of the Common Ancestor of Humans and Chimpanzees," *American Journal of Human Genetics* 68, no. 2 (2001): 444–456, https://doi.org/10.1086/318206.
7. Peter Godfrey-Smith, *Other Minds: The Octopus, the Sea, and the Deep Origins of Consciousness* (New York: Farrar, Straus and Giroux, 2016), 7.
8. Godfrey-Smith, *Other Minds*, 9.
9. Godfrey-Smith, *Other Minds*, 9.
10. For a contemporary account of animism that understands personhood and relationship more broadly than human or animal sentience, see Graham Harvey, *Animism: Respecting the Living World*, new ed. (London: Hurst & Company, 2017).
11. Marc Bekoff and Jessica Pierce, *Wild Justice: The Moral Lives of Animals* (Chicago: University of Chicago Press, 2009), 34–35.
12. Keith M. Kendrick et al., "Sheep Don't Forget a Face," *Nature* 414 (2001): 165–166, https://doi.org/10.1038/35102669.
13. Safina, *Beyond Words*, 260.
14. Godfrey-Smith, *Other Minds*, 50.
15. Safina, *Beyond Words*, 194–198.
16. Godfrey-Smith, *Other Minds*, 52, 56.
17. Safina, *Beyond Words*, 68–69.
18. Safina, *Beyond Words*, 238.
19. Peter Singer, *Writings on an Ethical Life* (New York: Ecco Press, 2000), xv–xvi.

4. Look and See

1. The original John Templeton Foundation project page, "Does the Universe Have a Purpose?" formerly at http://www.templeton.org/questions/purpose, is no longer available at that address; the URL now redirects to unrelated material.
2. Jim Holt, *Why Does the World Exist?: An Existential Detective Story* (New York: Liveright Publishing Corporation, 2012), 34.

3. The Holy Bible, New Revised Standard Version Updated Edition (NRSVue), Genesis 1:1–3.
4. Holt, *Why Does the World Exist?* 74–75.
5. Lawrence M. Krauss, *A Universe from Nothing: Why There Is Something Rather than Nothing* (New York: Free Press, 2012), 3.
6. Holt, *Why Does the World Exist?* 84.
7. Thomas S. Kuhn, *The Structure of Scientific Revolutions*, 4th ed., 50th anniversary ed., with an introductory essay by Ian Hacking (Chicago: University of Chicago Press, 2012), esp. chaps. 2, 6–7, and 9.
8. Ludwig Wittgenstein, *Philosophical Investigations*, trans. G. E. M. Anscombe, rev. P. M. S. Hacker and Joachim Schulte, 4th ed. (Malden, MA: Wiley-Blackwell, 2009), §66.

5. What Doesn't Go Away

1. Steven Weinberg, *To Explain the World: The Discovery of Modern Science* (New York: Harper, 2015), x.
2. Weinberg, *To Explain the World*, 4–6.
3. Weinberg, *To Explain the World*, 240.
4. Ibid.
5. Weinberg, *To Explain the World*, 16.
6. Weinberg, *To Explain the World*, 20.
7. Weinberg, *To Explain the World*, 33.
8. Weinberg, *To Explain the World*, 55.
9. Weinberg, *To Explain the World*, 64.
10. Weinberg, *To Explain the World*, 65.
11. Weinberg, *To Explain the World*, 146–148.
12. Weinberg, *To Explain the World*, 153.
13. Weinberg, *To Explain the World*, 156.
14. Weinberg, *To Explain the World*, 170.
15. Philip K. Dick, "How to Build a Universe That Doesn't Fall Apart Two Days Later," in *The Shifting Realities of Philip K. Dick: Selected Literary and Philosophical Writings*, ed. Lawrence Sutin (New York: Vintage Books, 1995), 261.
16. J. B. S. Haldane, *Possible Worlds and Other Essays* (London: Chatto & Windus, 1927), 286.
17. Weinberg, *To Explain the World*, 254.
18. Weinberg, *To Explain the World*, 42.
19. Weinberg, *To Explain the World*, 254–255.

6. The Sun Doesn't Rise

1. Robert Poole, *Earthrise: How Man First Saw the Earth* (New Haven: Yale University Press, 2008), 8–9.
2. Robert Poole, *Earthrise*, 8.
3. Carl Sagan, *Pale Blue Dot: A Vision of the Human Future in Space* (New York: Random House, 1994), 191.
4. David Christian, *Origin Story: A Big History of Everything* (New York: Little, Brown and Company, 2018), viii.
5. Pierre Teilhard de Chardin, "A Note on Progress," in *The Future of Man*, trans. Norman Denny (New York: Harper & Row, 1964), 34.
6. Christian, *Origin Story*, vii–xiii.
7. Michael Dowd, *Thank God for Evolution: How the Marriage of Science and Religion Will Transform Your Life and Our World* (New York: Viking, 2007), xii.

7. Stop Being One Person

1. David Foster Wallace, *This Is Water: Some Thoughts, Delivered on a Significant Occasion, about Living a Compassionate Life* (New York: Little, Brown and Company, 2009), 3–5.
2. Robert Kegan, *In Over Our Heads: The Mental Demands of Modern Life* (Cambridge, MA: Harvard University Press, 1994), 32–34.
3. Jeffrey J. Kripal et al., *Comparing Religions: The Study of Us That Changes Us*, 2nd ed. (Wiley-Blackwell, 2024), 465–466.
4. Kelly Weinersmith and Zach Weinersmith, *A City on Mars: Can We Settle Space, Should We Settle Space, and Have We Really Thought This Through?* (New York: Penguin Press, 2023), 32.
5. I share this as a minister, not a physician. Psychedelics carry genuine risks and are not right for everyone, particularly people with certain psychiatric histories or those taking medications that may interact dangerously with them. Their legal status also varies widely. Nothing here is medical or legal advice. Anyone drawn to these experiences should approach them only where legal and with appropriate preparation, care, and support.
6. Carl Sagan, *Cosmos* (New York: Random House, 1980), 4: "We are a way for the cosmos to know itself."
7. Our World in Data, "CO_2 Emissions per Capita," based on Global Carbon Project, *Global Carbon Budget*, https://ourworldindata.org/grapher/co-emissions-per-capita; World Bank, "GDP (Current US$)—

Germany," *World Development Indicators*, https://data.worldbank.org/indicator/NY.GDP.MKTP.CD?locations=DE.

8. Greta Thunberg, "1.5: Our Carbon Budget," in *The Climate Book: The Facts and the Solutions*, ed. Greta Thunberg (New York: Penguin Press, 2023), iii.
9. Thunberg, ed., *The Climate Book*, 393.
10. Thunberg, ed., *The Climate Book*, 4.
11. The Carbon Almanac Network, *Generation Carbon: It's Time to Start: A Carbon Almanac for Kids* (independently published, 2022), 31.
12. Rebecca Solnit and Thelma Young Lutunatabua, eds., *Not Too Late: Changing the Climate Story from Despair to Possibility* (Chicago: Haymarket Books, 2023), 7.
13. Kathleen Dean Moore, *Great Tide Rising: Towards Clarity and Moral Courage in a Time of Planetary Change* (Berkeley, CA: Counterpoint, 2016), 292.

8. Choosing Earth

1. Tim Peake, *Ask an Astronaut: My Guide to Life in Space* (New York: Little, Brown and Company, 2017), 240.
2. K. and Z. Weinersmith, *A City on Mars*, 22.
3. K. and Z. Weinersmith, *A City on Mars*, 2.
4. K. and Z. Weinersmith, *A City on Mars*, 11.
5. K. and Z. Weinersmith, *A City on Mars*, 137–138.
6. Adam Becker, *More Everything Forever: AI Overlords, Space Empires, and Silicon Valley's Crusade to Control the Fate of Humanity* (New York: Basic Books, 2025), 6.
7. Mary-Jane Rubenstein, *Astrotopia: The Dangerous Religion of the Corporate Space Race* (Chicago: University of Chicago Press, 2022), 18–22.
8. The aphorism "No matter where you go, there you are" crystallizes an older spiritual insight found, for example, in Thomas à Kempis's *The Imitation of Christ* (ca. 1440), where he writes, "wherever you go you are burdened with yourself." The modern proverbial form is attested at least as early as a 1955 student publication at Pennsylvania State University and was later popularized by *The Adventures of Buckaroo Banzai Across the 8th Dimension* (1984). Jon Kabat-Zinn's *Wherever You Go, There You Are: Mindfulness Meditation in Everyday Life* (1994) takes the proverb as its title and helped make it widely known in mindfulness and therapeutic circles. For more, see https://quoteinvestigator.com/2020/11/22/where-you-go/.
9. K. and Z. Weinersmith, *A City on Mars*, 382.

10. U.S. Department of Labor, Occupational Safety and Health Administration, "Federal Safety Inspections at Three Amazon Warehouse Facilities Find Company Exposed Workers to Ergonomic, Struck-by Hazards," January 18, 2023, https://www.osha.gov/news/newsreleases/osha-national-news-release/20230118.
11. Becker, *More Everything Forever*, 27–28.
12. David Remnick, "The Human Cost of DOGE's War on U.S.A.I.D.," *The New Yorker*, July 12, 2026, https://www.newyorker.com/news/the-new-yorker-interview/the-human-cost-of-doges-war-on-usaid; Nicholas Enrich, *Into the Wood Chipper: A Whistleblower's Account of How the Trump Administration Shredded USAID*, foreword by Atul Gawande (S&S/Summit Books, 2026).
13. World Meteorological Organization, *State of the Global Climate 2025* (Geneva: World Meteorological Organization, 2026), https://wmo.int/publication-series/state-of-global-climate/state-of-global-climate-2025; see also Julia Conley, "'Every Key Climate Indicator Is Flashing Red' in New UN Report," *Common Dreams*, March 2026, https://www.commondreams.org/news/un-report-climate.
14. Lisa Friedman and Harry Stevens, "How Lee Zeldin Shifted the Mission—and the Message—of the E.P.A.," *New York Times*, April 12, 2026, https://www.nytimes.com/interactive/2026/04/12/climate/lee-zeldin-epa-mission-language.html; Tom Perkins, "Trump to Revoke Protections for Endangered Species in Gulf of Mexico," *The Guardian*, March 30, 2026, https://www.theguardian.com/us-news/2026/mar/30/trump-protections-endangered-species-gulf-of-mexico.
15. Bill McKibben, *Here Comes the Sun: A Last Chance for the Climate and a Fresh Chance for Civilization* (New York: W. W. Norton, 2025), 3–4.
16. Tim Appenzeller, "Science's 2025 Breakthrough of the Year," *Science*, December 18, 2025, https://www.science.org/content/article/breakthrough-2025.
17. Rachel Abrams, David Gelles, and Brad Plumer, "The Future of Energy Has Arrived—Just Not in the U.S.," *The Daily*, *New York Times*, November 18, 2025, podcast, https://www.nytimes.com/2025/11/18/podcasts/the-daily/cop30-china-us-renewables-energy.html.
18. Questions about how transformative technologies should be governed for the common good extend well beyond climate change. Artificial intelligence raises them with particular urgency. If AI dramatically increases productivity while displacing human labor, who should own the wealth it creates? Various economists, policymakers, and technology thinkers have proposed treating AI as a public resource whose benefits should be broadly shared through mechanisms such as public

ownership, sovereign wealth funds, or universal dividends rather than accruing primarily to a handful of corporations and investors.

19. Paul Krugman, "Renewable Energy and National Security," Paul Krugman (Substack), March 6, 2026, https://paulkrugman.substack.com/p/renewable-energy-and-national-security.
20. Bill McKibben, "Sunlight Travels 93 Million Miles to Reach the Earth," *The Crucial Years* (Substack), March 6, 2026, https://billmckibben.substack.com/p/sunlight-travels-93-million-miles-63b.

9. Hidden in Plain Sight

1. Richard Higgins, *Thoreau and the Language of Trees* (Oakland: University of California Press, 2017), 2–3.
2. Higgins, *Thoreau and the Language of Trees*, 38.
3. Higgins, *Thoreau and the Language of Trees*, 91.
4. Higgins, *Thoreau and the Language of Trees*, 55.
5. Peter Wohlleben, *The Hidden Life of Trees: What They Feel, How They Communicate—Discoveries from a Secret World*, trans. Jane Billinghurst (Vancouver, BC: Greystone Books, 2016), vii.
6. Wohlleben, *The Hidden Life of Trees*, vii–viii.
7. Wohlleben, *The Hidden Life of Trees*, 65.
8. Wohlleben, *The Hidden Life of Trees*, viii.
9. Wohlleben, *The Hidden Life of Trees*, 82–83.
10. Wohlleben, *The Hidden Life of Trees*, 47–48.
11. Safina, *Beyond Words*, 24.
12. Unlike the approximately 1.2 percent human–chimpanzee difference and approximately 99.9 percent human–human identity figures used elsewhere in this book—each of which depends on a particular method of genomic comparison—there is no single meaningful percentage for how much DNA humans "share with trees." Estimates vary according to the species compared and whether researchers measure nucleotide identity, homologous genes, orthologs, protein-coding sequences, or other genomic features. The broader point is well established: humans and trees retain homologous genes, reflecting deep shared evolutionary ancestry. See Adrian M. Altenhoff et al., "The OMA Orthology Database in 2015: Function Predictions, Better Plant Support, Synteny View and Other Improvements," *Nucleic Acids Research* 43, no. D1 (2015): D240–D249; G. A. Tuskan et al., "The Genome of Black Cottonwood, *Populus trichocarpa* (Torr. & Gray)," *Science* 313, no. 5793 (2006): 1596–1604.
13. Wohlleben, *The Hidden Life of Trees*, x.

14. Doug Duvall, "Stumped by Giants: Woman Determines State's Largest Sugarberry Is in Frederick," *Frederick News-Post*, August 20, 2024, https://www.fredericknewspost.com/news/environment/stumped-by-giants-woman-determines-state-s-largest-sugarberry-is/article_d7a5b137-3465-5b08-b90b-dd64b1f068b6.html.
15. For a searchable registry of notable trees across the United States, see the National Champion Tree Program, https://nationalchampiontree.org.

10. Relationships All the Way Down

1. Ferris Jabr, "The Social Life of Forests," *The New York Times Magazine*, December 2, 2020, https://www.nytimes.com/interactive/2020/12/02/magazine/tree-communication-mycorrhiza.html.
2. Merlin Sheldrake, *Entangled Life: How Fungi Make Our Worlds, Change Our Minds & Shape Our Futures* (New York: Random House, 2020), 224.
3. John Muir, *My First Summer in the Sierra* (Boston: Houghton, Mifflin and Company, 1911), 110.
4. Paul Stamets, ed., *Fantastic Fungi: How Mushrooms Can Heal, Shift Consciousness, and Save the Planet* (Earth Aware Editions, 2019), 42.
5. Sheldrake, *Entangled Life*, 10.
6. Sheldrake, *Entangled Life*, 3–4.
7. Sheldrake, *Entangled Life*, 124–125.
8. Sheldrake, *Entangled Life*, 127.
9. Sheldrake, *Entangled Life*, 5.
10. Sheldrake, *Entangled Life*, 213.
11. Michael Pollan, *How to Change Your Mind: What the New Science of Psychedelics Teaches Us About Consciousness, Dying, Addiction, Depression, and Transcendence* (New York: Penguin Press, 2018), 219–49, 315–403.
12. Sheldrake, *Entangled Life*, 41.
13. Sheldrake, *Entangled Life*, 47.
14. Sheldrake, *Entangled Life*, 66.
15. Sheldrake, *Entangled Life*, 23.

11. Same Room, Different Worlds

1. Buddhist Geeks, "The Six Senses: Background," YouTube video, https://www.youtube.com/watch?v=MeQEVFA4fBo&t=157s.
2. Ed Yong, *An Immense World: How Animal Senses Reveal the Hidden Realms Around Us* (New York: Random House, 2022), 10.
3. Yong, *An Immense World*, 3–5.

4. Yong, *An Immense World*, 5–6.
5. Yong, *An Immense World*, 5.
6. Yong, *An Immense World*, 8.
7. Yong, *An Immense World*, 18–19.
8. Yong, *An Immense World*, 19–20.
9. Yong, *An Immense World*, 23.
10. Yong, *An Immense World*, 24–26.
11. Yong, *An Immense World*, 62–63.
12. Yong, *An Immense World*, 63.
13. Yong, *An Immense World*, 84–85, 88.
14. Yong, *An Immense World*, 86–87.
15. Yong, *An Immense World*, 88.
16. Marcel Proust, *The Prisoner*, trans. C. K. Scott Moncrieff (1929), quoted in Ed Yong, *An Immense World*, 16.
17. Yong, *An Immense World*, 11.
18. Yong, *An Immense World*, 354–355.

12. Before Allegiance

1. Dina Gilio-Whitaker, *As Long as Grass Grows: The Indigenous Fight for Environmental Justice, from Colonization to Standing Rock* (Boston: Beacon Press, 2019), 10.
2. Gilio-Whitaker, *As Long as Grass Grows*, ix.
3. Gilio-Whitaker, *As Long as Grass Grows*, 49. Gilio-Whitaker's estimate of a roughly 99 percent decline reflects the high end of a longstanding scholarly debate over the Indigenous population before European contact. Estimates vary widely because no reliable pre-contact census exists; lower population estimates would yield a decline closer to 90–95 percent. The catastrophic scale of the population collapse is not seriously disputed, although scholars continue to debate both its precise magnitude and how broadly the term genocide should be applied. Epidemic disease caused enormous mortality, while warfare, forced removal, dispossession, destruction of subsistence resources, and deliberate killing also contributed.
4. Robin Wall Kimmerer, *Braiding Sweetgrass: Indigenous Wisdom, Scientific Knowledge, and the Teachings of Plants* (Minneapolis: Milkweed Editions, 2013), 29.
5. Kimmerer, *Braiding Sweetgrass*, 40–41.
6. Kimmerer, *Braiding Sweetgrass*, 41.
7. Kimmerer, *Braiding Sweetgrass*, 44.
8. Kimmerer, *Braiding Sweetgrass*, 47.
9. Kimmerer, *Braiding Sweetgrass*, 106.

10. Kimmerer, *Braiding Sweetgrass*, 106.
11. Richard J. Ellis, *To the Flag: The Unlikely History of the Pledge of Allegiance* (Lawrence: University Press of Kansas, 2005), 17–21.
12. Kimmerer, *Braiding Sweetgrass*, 107–108.
13. Kimmerer, *Braiding Sweetgrass*, 111.

13. We Become What We Attend To

1. Gloria Mark, *Attention Span: A Groundbreaking Way to Restore Balance, Happiness and Productivity* (Hanover Square Press, 2023), 11.
2. Mark, *Attention Span*, 12.
3. Mark, *Attention Span*, 16.
4. Carey Nieuwhof, "3 Simple Steps to Get to Inbox Zero Today (And Why It Will Bring You More Peace Than You Think)," *CareyNieuwhof.com*, October 5, 2015, https://careynieuwhof.com/3-simple-steps-to-get-to-inbox-zero-today-and-why-it-will-bring-you-more-peace-than-you-think/.
5. Thich Nhat Hanh, *Nothing to Do, Nowhere to Go: Waking Up to Who You Are* (Berkeley, CA: Parallax Press, 2007).
6. Mark, *Attention Span*, 46.
7. Mark, *Attention Span*, 44–46.
8. Mark, *Attention Span*, 95–96.
9. Mark, *Attention Span*, 103–104.
10. Mark, *Attention Span*, 104.
11. Mark, *Attention Span*, 225.
12. Mark, *Attention Span*, 107.
13. Mark, *Attention Span*, 146.
14. Mark, *Attention Span*, 161.
15. Mark, *Attention Span*, 185.
16. Mark, *Attention Span*, 270.
17. Carey Nieuwhof, "Why They're Not Actually Your Friends (A Hard Truth for Pastors and Leaders)," *CareyNieuwhof.com*, October 17, 2021, https://careynieuwhof.com/why-theyre-not-actually-your-friends-a-hard-truth-for-pastors-and-leaders/.
18. Ibid. Carey Nieuwhof explores these ideas further in *At Your Best: How to Get Time, Energy, and Priorities Working in Your Favor.*
19. Jenny Odell, *How to Do Nothing: Resisting the Attention Economy* (Brooklyn, NY: Melville House, 2019), xviii.

14. Never Say Never, Never Say Always

1. Alan Levinovitz, *Natural: How Faith in Nature's Goodness Leads to Harmful Fads, Unjust Laws, and Flawed Science* (Boston: Beacon Press, 2020), 8.
2. The frequently cited figure of ~99.9% identity reflects differences at individual DNA-letter positions. Broader measures that also include larger genomic variants yield a figure closer to ~99.6% identity. These figures are not contradictory; they measure genomic variation in different ways. Francis S. Collins, "The Human Genome Project: Revealing the Shared Inheritance of All Humankind," *Cancer* 91, no. 1 suppl. (2001): 221–225, https://pubmed.ncbi.nlm.nih.gov/11148583/; National Human Genome Research Institute, "Human Genomic Variation," https://www.genome.gov/about-genomics/educational-resources/fact-sheets/human-genomic-variation.
3. Levinovitz, *Natural*, 6.
4. Levinovitz, *Natural*, 19.
5. Levinovitz, *Natural*, 22.
6. Levinovitz, *Natural*, 23.
7. Barry Johnson, *Polarity Management: Identifying and Managing Unsolvable Problems* (Amherst, MA: HRD Press, 2014).
8. Teddy Rosenbluth, "Measles Cases Hit New Record in U.S., as Vaccination Rates Wane," *New York Times*, July 24, 2026, https://www.nytimes.com/2026/07/24/well/measles-record-united-states-numbers.html; Centers for Disease Control and Prevention, "Measles Cases and Outbreaks," updated August 14, 2026, https://www.cdc.gov/measles/data-research/index.html; Centers for Disease Control and Prevention, "Healthcare Providers: Stay Alert for Measles Cases," January 26, 2026, https://www.cdc.gov/measles/hcp/clinical-overview/stay-alert-for-measles-cases.html.

15. Two Weeks on a Picnic Table

1. Cal Newport, *Slow Productivity: The Lost Art of Accomplishment Without Burnout* (Portfolio, 2024), 1.
2. Newport, *Slow Productivity*, 2.
3. Newport, *Slow Productivity*, 7.
4. Newport, *Slow Productivity*, 22.
5. Newport, *Slow Productivity*, 13.
6. Newport, *Slow Productivity*, 13–14.
7. Newport, *Slow Productivity*, 28–29.

8. Newport, *Slow Productivity*, 31–35.
9. Newport, *Slow Productivity*, 8.
10. Newport, *Slow Productivity*, 41.
11. Newport, *Slow Productivity*, 53.
12. Newport, *Slow Productivity*, 176.
13. Newport, *Slow Productivity*, 47.
14. Newport, *Slow Productivity*, 48.
15. Newport, *Slow Productivity*, 53.
16. Newport, *Slow Productivity*, 52.
17. Newport, *Slow Productivity*, 111.
18. Newport, *Slow Productivity*, 112.
19. Newport, *Slow Productivity*, 153.
20. Newport, *Slow Productivity*, 115.
21. Newport, *Slow Productivity*, 147.
22. Newport, *Slow Productivity*, 148.
23. Newport, *Slow Productivity*, 149.
24. Newport, *Slow Productivity*, 149.
25. Newport, *Slow Productivity*, 173.
26. Newport, *Slow Productivity*, 169.
27. Newport, *Slow Productivity*, 170–172.
28. Newport, *Slow Productivity*, 216.
29. Newport, *Slow Productivity*, 217.
30. Newport, *Slow Productivity*, 219–220.

16. Coming to Our Senses

1. Jon Kabat-Zinn, *Full Catastrophe Living: Using the Wisdom of Your Body and Mind to Face Stress, Pain, and Illness*, rev. ed. (New York: Bantam Books, 2013), 3–4.
2. Kenneth Folk, *Contemplative Fitness* (self-published manuscript, 2013), introduction, https://www.dharmaoverground.org/documents/portlet_file_entry/10128/KennethFolkContemplativefitness.pdf/abccaba2-42b6-4f86-90c1-bf2646c2ea02?download=true.
3. Kristin Neff and Christopher Germer, *The Mindful Self-Compassion Workbook: A Proven Way to Accept Yourself, Build Inner Strength, and Thrive* (New York: Guilford Press, 2018).
4. Christine Caldwell, *Bodyfulness: Somatic Practices for Presence, Empowerment, and Waking Up in This Life* (Boston: Shambhala Publications, 2018).
5. Reginald A. Ray, *Somatic Descent: How to Unlock the Deepest Wisdom of the Body* (Boulder, CO: Shambhala, 2020), 22.
6. Ray, *Somatic Descent*, 48.

7. Sonya Renee Taylor, *The Body Is Not an Apology: The Power of Radical Self-Love*, 2nd ed. (Oakland, CA: Berrett-Koehler Publishers, 2021), 44.
8. Fortune Business Insights, Cosmetics Market Size, Share, Growth, and Industry Report, 2034, report, July 17, 2026, https://www.fortunebusinessinsights.com/cosmetics-market-102614.
9. Taylor, *The Body Is Not an Apology*, xiv–xv.
10. Taylor, *The Body Is Not an Apology*, 10.
11. Taylor, *The Body Is Not an Apology*, 21.
12. Taylor, *The Body Is Not an Apology*, 22.
13. Taylor, *The Body Is Not an Apology*, 23.
14. Taylor, *The Body Is Not an Apology*, 25.
15. "You don't have to believe everything you think" is a widely used maxim in contemporary mindfulness and cognitive-behavioral teaching. Its precise origin is unclear, but the phrase captures a central insight of both traditions: thoughts can be noticed without automatically being treated as facts.
16. Taylor, *The Body Is Not an Apology*, 26–27.

IV. Becoming Good Ancestors

1. Joanna Macy and Chris Johnstone, *Active Hope: How to Face the Mess We're In with Unexpected Resilience and Creative Power*, rev. ed. (Novato, CA: New World Library, 2022).

17. Communities Worth Choosing

1. Reuters, "Pictures of the Year: Religion," December 19, 2025, https://www.reuters.com/pictures/pictures-year-religion-2025-12-11/. These snapshots were inspired by Reuters' year-end photography on religion and supplemented with reporting from Reuters, the Associated Press, and other major news organizations to illustrate a representative year in global religious life.
2. Frances Mao, Stephanie Convery, and Krishani Dhanji, "Bondi Attack: Vigils Held as More Than A$1m Raised for Man Who Tackled Shooter —As It Happened," *The Guardian*, updated December 15, 2025, https://www.theguardian.com/australia-news/live/2025/dec/14/bondi-beach-shooting-live-updates-victims-terror-attack-jewish-festival-hanukah.
3. Becka A. Alper and Aleksandra Sandstrom, "If the U.S. Had 100 People: Charting Americans' Religious Affiliations," *Pew Research Center*, November 13, 2025, https://www.pewresearch.org/short-reads/2025/11/13/if-the-u-s-had-100-people-charting-americans-religious-affiliations/.

Figures are rounded to whole persons and therefore do not sum exactly to 100; the source also omits the 1 percent who did not answer.

4. Unitarian Universalist Association, *2025 Certification Report*, https://secure.uua.org/certification-report/; Ryan Burge, "What About the Unitarians? What the Data Actually Says About Unitarian Universalism in America," *Graphs About Religion*, May 28, 2026, https://www.graphsaboutreligion.com/p/what-about-the-unitarians. At that scale, Unitarian Universalists would not even make up one full person—only about one-tenth of a person in a town of 100.
5. Ryan P. Burge, *The American Religious Landscape: Facts, Trends, and the Future* (Oxford University Press, 2025), 31–32.
6. Burge, *The American Religious Landscape*, 52, 57–59, 70.
7. Ryan P. Burge, "The Aftermath of the Schism in the United Methodist Church," *Graphs About Religion* (Substack), April 25, 2024, https://www.graphsaboutreligion.com/p/the-aftermath-of-the-schism-in-the.
8. Ryan P. Burge, *The Nones: Where They Came From, Who They Are, and Where They Are Going*, 2nd ed. (Minneapolis: Fortress Press, 2023).
9. Gregory A. Smith et al., "Decline of Christianity in the U.S. Has Slowed, May Have Leveled Off," *Pew Research Center*, February 26, 2025, https://www.pewresearch.org/religion/2025/02/26/decline-of-christianity-in-the-us-has-slowed-may-have-leveled-off/.
10. Ryan P. Burge, *The Nones*.
11. Ryan P. Burge, "The Four Types of Nones," *Graphs About Religion* (Substack), August 9, 2025, https://www.graphsaboutreligion.com/p/the-four-types-of-nones. The percentages for these four groups are rounded to the nearest whole number and therefore total 101% rather than exactly 100%.
12. Ryan P. Burge, "The Nones Project: NiNos," *Graphs About Religion* (Substack), September 13, 2025, https://www.graphsaboutreligion.com/p/the-nones-project-ninos; Ryan P. Burge and Tony Jones, "Nones in Name Only," *The Nones Project*, https://www.thenonesproject.com/nones-in-name-only.
13. Ryan P. Burge and Tony Jones, "Spiritual but Not Religious (SBNR)," *The Nones Project*, https://www.thenonesproject.com/sbnr.
14. Ryan P. Burge, "The Nones Project: Dones," *Graphs About Religion* (Substack), October 11, 2025, https://www.graphsaboutreligion.com/p/the-nones-project-dones; Ryan P. Burge and Tony Jones, "Religious Disengageds," *The Nones Project*, https://www.thenonesproject.com/disengageds.
15. Ryan P. Burge, "The Nones Project: Zealous Atheists," *Graphs About Religion* (Substack), October 25, 2025, https://www.graphsaboutreligion.com/p/the-nones-project-zealous-atheists.

16. Burge, *The American Religious Landscape*, 251–252.
17. Burge, *The American Religious Landscape*, 253–254.
18. Burge, *The American Religious Landscape*, 254.
19. Burge, *The American Religious Landscape*, 257–259.
20. Burge, *The American Religious Landscape*, 259.

18. How to Be a Good Ancestor

1. The first and second clauses echo Deuteronomy 6:11 and a nineteenth-century proverb about planting trees under whose shade one will never sit; the third clause is a modern paraphrase rather than a traceable quotation. See Deuteronomy 6:11; and for the proverb's history, see *Quote Investigator*, "Blessed Are Those Who Plant Trees Under Whose Shade They Will Never Sit," https://quoteinvestigator.com/2020/04/29/tree-shade/.
2. The Rev. Joanna Fontaine Crawford, "The Essay I Won't Be Sending the RSCC," *The Journey* (blog), March 31, 2010, https://uuminister.blogspot.com/2010/03/essay-i-wont-be-sending-rscc.html.
3. Richard Rorty, *The Rorty Reader*, ed. Christopher J. Voparil and Richard J. Bernstein (Malden, MA: Wiley-Blackwell, 2010), 85.
4. The figure 4,096 refers to positions in a family tree, not necessarily 4,096 different people. As a family tree expands backward, separate branches often lead to the same ancestor—for example, when two distant cousins share an earlier ancestor. The repeated appearance of one person in multiple positions is known as pedigree collapse.
5. Icurra, "Big Picture, Big Questions," *Medium*, March 20, 2020, https://icurra.medium.com/big-picture-big-questions-79e4f08e60af.
6. Roman Krznaric, *The Good Ancestor: A Radical Prescription for Long-Term Thinking* (New York: The Experiment, 2020), 3.
7. Krznaric, *The Good Ancestor*, 6, 12.
8. Krznaric, *The Good Ancestor*, 34.
9. Krznaric, *The Good Ancestor*, 49.
10. Krznaric, *The Good Ancestor*, 57.
11. Krznaric, *The Good Ancestor*, 69, 71. "We do not inherit the Earth from our ancestors; we borrow it from our children." This saying is of uncertain origin and has been misattributed to Chief Seattle, Ralph Waldo Emerson, and others without evidence. The earliest close textual parallel appears in Wendell Berry, *The Unforeseen Wilderness: An Essay on Kentucky's Red River Gorge* (Lexington: University Press of Kentucky, 1971), 26: "A man who knows that the world is not given by his fathers, but borrowed from his children." For a thorough genealogy of the saying, see Garson O'Toole, "We Do Not Inherit the Earth from Our

Ancestors; We Borrow It from Our Children," *Quote Investigator*, January 22, 2013, https://quoteinvestigator.com/2013/01/22/borrow-earth/.

12. Krznaric, *The Good Ancestor*, 71.
13. Krznaric, *The Good Ancestor*, 113.
14. Krznaric, *The Good Ancestor*, 243.
15. Thích Nhất Hạnh, *Zen and the Art of Saving the Planet* (New York: HarperOne, 2021), vii.
16. Nhất Hạnh, *Zen and the Art of Saving the Planet*, viii.
17. Andreas Malm, *How to Blow Up a Pipeline: Learning to Fight in a World on Fire* (London: Verso, 2021), 20.
18. Malm, *How to Blow Up a Pipeline*, 102.
19. Malm, *How to Blow Up a Pipeline*, 134.

19. Because I'm in Love

1. David Gelles, "Scientists Are Freaking Out About Ocean Temperatures," *New York Times*, February 27, 2024, https://www.nytimes.com/2024/02/27/climate/scientists-are-freaking-out-about-ocean-temperatures.html.
2. Gelles, "Scientists Are Freaking Out About Ocean Temperatures."
3. Shannon Osaka and Sarah Kaplan, "Before the Floods, Asheville Was Called a 'Climate Haven.' Is Anywhere Safe?" *Washington Post*, October 1, 2024, https://www.washingtonpost.com/climate-environment/2024/10/01/asheville-climate-haven-flooding-hurricane/.
4. Wendell Berry, endorsement statement for Charles E. Little, *The Dying of the Trees: The Pandemic in America's Forests* (New York: Penguin Books, 1997).
5. Robert Macfarlane, introduction to *[To] the Last [Be] Human*, by Jorie Graham (Port Townsend, WA: Copper Canyon Press, 2022), xvi.
6. Jameson's earlier formulation is more precise: "It seems to be easier for us today to imagine the thoroughgoing deterioration of the earth and of nature than the breakdown of late capitalism; perhaps that is due to some weakness in our imaginations." Fredric Jameson, *The Seeds of Time* (New York: Columbia University Press, 1994), xii. In "Future City," he invokes the now-familiar, but qualified, version: "Someone once said that it is easier to imagine the end of the world than to imagine the end of capitalism." Fredric Jameson, "Future City," *New Left Review* 21 (May–June 2003): 76.
7. Ursula K. Le Guin, "National Book Foundation Medal: Acceptance Speech," National Book Awards, New York, November 19, 2014, Ursula K. Le Guin official website, https://www.ursulakleguin.com/nbf-medal.

8. Theodore Parker, "Of Justice and the Conscience," in *Ten Sermons of Religion* (Boston: Rufus Leighton, 1853): "I do not pretend to understand the moral universe, the arc is a long one, my eye reaches but little ways. I cannot calculate the curve and complete the figure by the experience of sight; I can divine it by conscience. But from what I see I am sure it bends towards justice." Martin Luther King Jr. shortened that quote to "the arc of the moral universe is long, but it bends toward justice."
9. Brian D. McLaren, *Life After Doom: Wisdom and Courage for a World Falling Apart* (New York: St. Martin's Essentials, 2024), 83.
10. ecoAmerica, "5 Steps to Effective Climate Communication," *Climate Action Guides*, https://ecoamerica.org/climate-action-sheet/5-steps-to-effective-climate-communication/.

20. Apocaloptimism

1. Benjamin Storrow and Jean Chemnick, "How Trump Gutted Climate Policy in 30 Days," *Politico*, February 22, 2025, https://www.politico.com/news/2025/02/21/trump-gutted-climate-policy-across-federal-government-00204998; Ella Nilsen, "Trump Takes an Ax to More Than a Dozen Pollution Rules in Rapid-Fire Deregulation," *CNN*, March 12, 2025, https://www.cnn.com/2025/03/12/climate/trump-ev-power-plant-rollbacks/index.html.
2. Coral Davenport, Christopher Flavelle, and Lisa Friedman, "What Trump's Environmental Record Says About a Second Term," *New York Times*, November 1, 2024, https://www.nytimes.com/2024/11/01/climate/trump-environment-election.html.
3. Gus Speth, "A New Consciousness and the Eight-fold Way towards Sustainability," interview by Miriam Vilela, *Earth Charter International Podcast*, January 13, 2021, https://earthcharter.org/podcasts/gus-speth/.
4. Ram Dass, quoted in "Apocaloptimistic Dharma," Rocky Mountain Ecodharma Retreat Center, May 5–11, 2025, https://rmerc.org/events/apocaloptimistic-dharma/.
5. Ezra Klein and Derek Thompson, *Abundance* (New York: Avid Reader Press, 2025), 95.
6. Klein and Thompson, *Abundance*, 96.